Introduction to the Philosophy of Educational Research

RIVER PUBLISHERS SERIES IN INNOVATION AND CHANGE IN EDUCATION - CROSS-CULTURAL PERSPECTIVE

Series Editors:

Xiangyun Du
Department of Learning and Philosophy, Aalborg University, Denmark
and
College of Education, Qatar University, Qatar

Nowadays, educational institutions are being challenged as professional competences and expertise become progressively more complex. This is mainly because problems are more technology-bounded, unstable and ill-defined with the involvement of various integrated issues. Solving these problems requires interdisciplinary knowledge, collaboration skills, and innovative thinking, among other competences. In order to facilitate students with the competences expected in their future professions, educational institutions worldwide are implementing innovations and changes in many respects.

This book series includes a list of research projects that document innovation and change in education. The topics range from organizational change, curriculum design and innovation, and pedagogy development to the role of teaching staff in the change process, students' performance in the areas of not only academic scores, but also learning processes and skills development such as problem solving creativity, communication, and quality issues, among others. An inter- or cross-cultural perspective is studied in this book series that includes three layers. First, research contexts in these books include different countries/regions with various educational traditions, systems.. and societal backgrounds in a global context. Second, the impact of professional and institutional cultures such as language, engineering, medicine and health, and teachers' education are also taken into consideration in these research projects. The third layer incorpotates individual beliefs, perceptions, identity development and skills development in the learning processes, and inter-personal interaction and communication within the cultural contexts in the first two layers.

We strongly encourage you as an expert within this field to contribute with your research and help create an international awareness of this scientific subject.

For a list of other books in this series, visit www.riverpublishers.com

Introduction to the Philosophy of Educational Research

Omar A. Ponce
Ana G. Méndez University, Cupey Campus, Puerto Rico, USA

José Gómez Galán
University of Extremadura, Spain, and Ana G. Méndez University, Cupey Campus, Puerto Rico, USA

Nellie Pagán-Maldonado
Ana G. Méndez University, Cupey Campus, Puerto Rico, USA

Angel L. Canales Encarnación
Ana G. Méndez University, Cupey Campus, Puerto Rico, USA

River Publishers

Routledge
Taylor & Francis Group

LONDON AND NEW YORK

Published 2021 by River Publishers
River Publishers
Alsbjergvej 10, 9260 Gistrup, Denmark
www.riverpublishers.com

Distributed exclusively by Routledge
4 Park Square, Milton Park, Abingdon, Oxon OX14 4RN
605 Third Avenue, New York, NY 10158

First published in paperback 2024

Introduction to the Philosophy of Educational Research / by Omar A. Ponce, José Gómez Galán, Nellie Pagán-Maldonado, Angel L. Canales Encarnación.

Routledge is an imprint of the Taylor & Francis Group, an informa business

Publisher's Note
The publisher has gone to great lengths to ensure the quality of this reprint but points out that some imperfections in the original copies may be apparent.

While every effort is made to provide dependable information, the publisher, authors, and editors cannot be held responsible for any errors or omissions.

ISBN: 978-87-7022-637-0 (hbk)
ISBN: 978-87-7004-293-2 (pbk)
ISBN: 978-1-003-33869-7 (ebk)

DOI: 10.1201/9781003338697

Contents

Preface

Education is an essential component of public policy in many countries around the world. Education is an instrument of social and economic development. From ancient Greece to the agricultural, then industrial, and now technological societies in which we live, education is essential in training new generations of world citizens. Each new generation needs to be equipped with the tools that will enable them to face the realities of life they inherit from past generations and the new ones they create with their visions of life. Through education, each generation is expected to solve and bring about the personal, social, and economic development it has been given to live. Researching education to understand its effects on individuals and society and make it more compelling is the foundation of educational research. This book, **Introduction to the Philosophy of Educational Research**, presents a historical−philosophical look at the development of educational research methods and how this has influenced their scientific effectiveness. The book is organized into 11 chapters to place in context the theory and practice of educational research, its origin, and development as a scientific research discipline, and the historical events that define its methodological evolution:

1. **Educational Research: Theory and Practice** (Chapter 1). It provides an overview of contemporary educational research.
2. **The Problems of Public Education and the Need to Investigate Education to Improve It** (Chapter 2). The origin of public education, its complexity, and the issues it has faced are examined. Solving the problems of public education awakens the interest in researching education and, thus, the origin of educational research.
3. **Philosophy-Based Research** (Chapter 3). The entry and development of quantitative and qualitative research models to educational research that defines academic research to the present are examined.
4. **The Quantitative and Qualitative Paradigm Debate: Alternate Visions of Research Education** (Chapter 4). It provides an analysis of the early academic and policy discussion on the strengths and

shortcomings of quantitative and qualitative research models as applied to educational research.

5. **The Emergence of Mixed Methods in Educational Research** (Chapter 5). The emergence of mixed methods research as a response to quantitative and qualitative educational research shortcomings is discussed.

6. **Science-Based Research to Provoke Evidence-Based Professional Practice: Political Consensus on How Educational Research Should Be** (Chapter 6). It discusses the second academic and political debate on educational research problems and the requirements that any educational research seeking public funding must satisfy.

7. **Education as an Object of Scientific Research** (Chapter 7). Education is known more from a philosophical than from a scientific point of view. Contemporary approaches to understanding education as an object of scientific inquiry and their implications for the development of educational research are examined.

8. **From Research Applied to Practice to Research Embedded in Practice** (Chapter 8). It examines the argument generated about the need to develop a model of educational research responsive to the particular phenomena of education.

9. **Educational Research in an Era of Ethics in Scientific Research** (Chapter 9). It examines academic study and the methodological and scientific effectiveness challenges it faces in meeting the requirements of Institutional Compliance Committees.

10. **The Used of Standarized to Measure Learning** (Chapter 10). It provided an overview look at the used of standarided test in education to measure learning as a way to established the educational effectiveness.

11. **Research-evaluation of learning.** In the same way the realiability of standadared test to measure students learning have becomes an issue, more educational researcher are using research methods to evaluate the efectiveness of educational programs and schools.

Each event is used to reflect on the effect it has had on the methodological development of contemporary educational research and its scientific effectiveness. We understand that this reading represents a way of introducing the reader to educational research. Without many technicalities, the reader is exposed to the development of scholarly research in its philosophical concepts or visions and its scientific research methodologies.

Each historical event provides the context to explain the origin and conceptual and methodological development of educational research.

The book is oriented to education students who are new to educational research. In some universities, training in academic research for new generations of educators and educational researchers begins at the bachelor's level and in master's programs. The book can be a topic of discussion in courses on educational research methods, educational philosophy, and educational polemics. We also believe that the book can be of value to university professors in the social sciences, humanities, natural sciences, and nursing who sometimes serve as members of doctoral dissertation and master's thesis committees in education. We hope that this book will be a valuable tool for developing educational research for its readers.

Omar A. Ponce, Ph.D.
José Gómez Galán, Ph.D., Ph.D.
Nellie Pagán-Maldonado, Ph.D.
Ángel Canales Encarnación, Ed.D.
Ana G. Méndez University
Cupey Campus

1

Educational Research: Theory and Practice

Omar A. Ponce

Since its beginning in the early 20th century, educational research has gradually developed as an academic discipline to understand and improve education. The complexity and dynamism of education have made it evident that educational research is much more than research methods. For this reason, the philosophical study of educational research emerges as a pillar in the development of educational research. At the beginning of the 21st century, the value and necessity of effective educational research are reinforced for four reasons. (a) Education remains an essential instrument for social and economic development in public policy in many countries worldwide. The educational needs of the 21st century may resemble, but are not the same as, the educational needs of industrial society. For example, student learning, student dropout, and student motivation have been common historical themes when discussing education. Still, the causes and contexts are completely different between the realities of industrial society and the realities of technological society. In other words, education is not a static, stable, or uniform reality. (b) Implementing the ideal of public education continues to pose problems and challenges in the realities of the technological society in which we live. The quest for quality public education has been and continues to be a battleground where educators, administrators, educational researchers, and policy makers clash. The political world continues to make changes in educational systems through legislation and educational policies, some scientifically based and others not. This is reflected in educational projects with names like learning assessment and accountability that create complicated dynamics for educators, students, and parents. (c) From the 1980s to the present, the logic on which current educational research methods solve academic problems has been questioned. The scope and effectiveness of educational research methods are questioned when examining the complexity and dynamism of education as an object of scientific research. In this sense, educational research needs to develop its research model that responds to the realities of education. The challenge is there, and the opportunities are there to make educational research an effective instrument for developing education in the society. Understanding these challenges and opportunities is possible if one understands the methodological evolutions that educational research has undergone and the reasons for these evolutions. It is the central theme and the main objective of this book. This chapter examines educational research as an academic discipline and a field of work. It provides the context for the discussion of the events that define the methodological evolution of educational research from its beginning to the time of writing this book. The chapter is organized into two sections to present the concept and context of 21st-century educational research: Theory and Practice.

Theory

There are many definitions of educational research (Hedges and Hanis-Martin, 2009; Condliffe and Shulman, 1999). The dominant position is to define it as "educational research." From this perspective, educational research means investigating educational practices, the effect of these practices on learning, and the study of academic problems (Johannigmeier & Richardson, 2008; McMillan, J. H. & Schumacher, S., 2005; Condliffe & Schulman, 1999; Segovia, J., 1997; Charles, C.M., 1988; Cohen & Manion, 1980). The interest in researching education emerges at the beginning of the 20th century, with the emergence of public education worldwide and the desire to develop it scientifically (Walters, 2009; Johannigmeier and Richardson, 2008; Condliffe, 2000). The argument at that time, and which is sustained at the beginning of the 21st century, was that scientific research could improve public education as it had happened in other professions. This interest emerges the term educational research to refer to research in public education issues (Johannigmeier and Richardson, 2008; Shavelson and Towne, 2002; Condliffe, 2000; Cohen and Manion, 1980).

From its beginning in the early 20th century to the beginning of the 21st century, the goal of educational research has been to make education an effective profession in training new generations of citizens and professionals (Ponce, 2014a; Barnhouse, Lareau and Ranis, 2009; Latorre, 2008; Elliot, 2007; Labarre, 2004; Carr and Kemmis, 1985). Historically, the discussion on the effectiveness of education has been related to the effect of teaching on learning, the nature of learning, the cost benefits of schooling and sustaining a public education system, the role of schooling in reducing poverty, and how instruction can improve a country's productivity and economic competitiveness (Johanningmeier and Richardson, 2008; Cohen and Barnes, 1999). For education to be effective, educational research must accomplish the following: (a) To generate the knowledge that the profession needs to develop (Ponce, 2014 and 2016) in the particularities of each school or educational system (Latorre, 2008); (b) To generate what works and does not work in schools to improve their effectiveness (Labaree, 2004); (c) To expose the conditions for developing a coherent and effective teaching–learning process in the various educational contexts that exist (Latorre, 2008; Eliott, 2007; Carr and Kemmis, 1985); (d) To identify the strategies used by educators in the various educational scenarios and validate these practices— this involves studying their beliefs about education to expose and channel them (Latorre, 2008; Pring, 2000; Carr and Kemmis, 1985); (e) To study educational objectives, teaching practices, and these practices' products to improve them if necessary (Biesta and Burbules, 2003; Pring, 2000); (f) To evaluate the scope of educational policies and determine whether they achieve their purpose (Biesta and Burbules, 2003; Pring, 2000; Ponce, Pagán-Maldonado and Gómez Galán, 2017, 2018a); (g) To generate educational theories that inform the practice of the profession and the development of educational policies (Shavelson and Towne, 2002; Pring, 2000; Carr and Kemmis, 1985); (h) it should not be limited to the study of problems in education, nor should it focus on the study of causal relationships between particular classroom activities with the outcomes pre-established in standards because its purpose is to make the field of education an effective and responsive profession to the needs of society (Elliott, 2007). Some authors believe that academic research is more effective in theory than in practice. Educational research has not been effective in practice because it has failed to achieve any of its objectives (Ponce, Pagán-Maldonado, & Gómez Galán, 2017, 2018). On the contrary, research has been plagued by controversies about how educational research should be to be effective (Ponce, 2016; Ponce, Pagán-Maldonado, & Gómez Galán, 2018).

Practice

In this early 21st century, educational research is developing and defining itself as a unique and self-deserving field of scientific inquiry (Koichiro, 2013). As a field of scientific inquiry, it seeks to expand its theories that effectively guide the practice of education (Green, 2010; Ponce, 2017). In the reality of education for a global and technological era, educational research seeks to transcend schooling research because of the impact education has on society (Lee, 2010). In the early 21st century, academic research also encompasses the study of K-12 education and university education in the development of society (Gomez Galán, 2015). There is much expectation from politicians, administrators, entrepreneurs, and international educational organizations that educational research will generate the knowledge to develop the educational practices and policies necessary to guarantee success in the implementation of the scientific-educational model (Pring, 2007; Ponce, Pagán-Maldonado, & Gómez Galán, 2017 & 2018). At this historical moment, educational research is incorporating technology into its scientific endeavors. Technology generates new data and understandings (Smeyers & Depaepe, 2008; Rice & Vastola, 2009; Gómez Galán, 2017). The core question that contemporary educational researchers are asking about education is what knowledge can be generated from this field (Green, 2010) and what type of research is helpful for these purposes (Gil Cantero and Reyero, 2014; Ponce, Pagán-Maldonado and Gómez Galán, 2017 and 2018).

In many countries, educational research is an institutionalized activity because its importance in the development of society is recognized (Condliffe, 2000; Pring, 2000; Shavelson and Towne, 2002; Johannigmeier and Richardson, 2008; Walters, Lareau and Ranis, 2009). For example, countries such as the United States, England, and Australia invest large amounts of money in educational research to improve public education and eradicate its problems. From their institutionalization emerge the influences that define their scientific culture. Following are the structures that define the field of educational research (Ponce, 2016; Ponce, Pagán-Maldonado and Gómez Galán, 2017 and 2018):

> **Government regulatory offices.** In many countries, government departments or divisions define the research needs of public education, regulate the allocation of money for educational research, and establish the quality criteria that academic research seeking public funding must meet. Some of these departments/divisions have institutes dedicated to researching and documenting the effectiveness of K-16 public education and issue management reports that influence the country's educational policy.

> **Research offices and divisions in education systems.** In universities and some primary and secondary school systems, educational research structures are identified by various names: offices of statistics and research, institutional assessment and research, or planning and

research. These structures handle multiple and complex responsibilities: (a) they conduct research and evaluate various aspects of the educational institution to inform management decision-making, (b) they generate the data and statistics to fulfill reports to licensing and accrediting agencies, (c) they establish policies to regulate the conduct of research in their educational institutions, or (d) they establish regulations that their researchers must satisfy in the procedures and ethics of their studies to safeguard the accountability of the institution they represent.

Professional associations of educational research. The institutionalization of scholarly research can also be seen in the presence of professional organizations of academic researchers which are formed for their professional development. These groups organize forums where research is presented, research advances are discussed, controversial issues are debated in practice, and policy recommendations are formulated to government agencies to regulate the practice of educational research.

Professional journals of educational research. Professional journals represent another manifestation of the institutionalization of scholarly research. They establish the platform for disseminating academic research and defining the criteria for publication and the quality that they accept as requirements for these purposes.

Universities. The vast majority of contemporary educational research is generated in universities. The bulk of educational research is generated in schools of education. Most of this research is oriented to advance knowledge of the various specialties of education (e.g., educational administration, physical education, and preschool education) and generate theories and models that guide practice by scientifically proven principles. There are also university professors in other academic disciplines who research public education through government-funded projects. The goal of government-funded educational research is to solve problems in public education (Walters, 2009). In this development, two philosophical positions on the role of educational research are visible. (a) **The instrumental view of educational research or applied research.** This position is known as educational research (Educational Research). It is research on educational practices and their policies. It is research to be applied to the practice of education (Phillips, 2005; Green, 2010; Clark, 2011; Gil Cantero and Reyero, 2014). This vision emanates from the positivist philosophy of scientific research.

The function of research is to improve the world by developing new and more effective techniques to replace traditional ones and the constant demonstration of how things work and can be improved (Hammersley, 2007). It is also known as applied research (Shavelson and Towne, 2002). (b) **The informational view of educational research or basic research.** This position is known as research in education (research in education − education research). It is social or empirical research on education without seeking practical applications. It does not pursue the interest of connecting to the practice of the profession and can use the research methods it understands. It is also known as basic and academic research (Clark, 2011). Research as an exercise to reflect on social and educational practices beyond its instrumental nature. Research is presented as the search for knowledge, an exercise of intellectual curiosity, and a critical exercise of evaluating current practices, even if there is satisfaction with the status quo of things (Hammersley, 2007).

Consulting offices and non-university organizations. Outside the university environment, private offices are dedicated to consulting and research-evaluation of educational programs subsidized by federal funds. In the United States of America, federally funded educational research-evaluation has become an industry (Walters, 2009).

Objectives of Educational Research

The focus of educational research is the education that occurs in educational systems and its effectiveness. The field of education revolves around five major areas that determine its effectiveness: (a) educational practices for teaching students, (b) student learning, (c) the social dynamics of education, (d) the problems of education, and (e) the management of educational institutions to channel teacher management toward the achievement of educational goals and objectives. Let us examine these areas to enter into the work of educational research and the distinctive that differentiate it from other forms of scientific research (Carr and Kemmis, 1985; Pring, 2000; Elliots, 2007; Ponce, 2014b; Ponce, 2016; Ponce, Pagán-Maldonado and Gómez Galán, 2017 and 2018):

Research on educational practices. The term educational practices are used to describe all the efforts, activities, and dynamics generated to provoke student learning. Educational practices are developed around the following components:

Educational goals and objectives. Educational goals and objectives constitute the guidelines to be followed in primary and secondary education. In universities, statements of vision, mission, and values are used. Educational research has studied the processes that have been observed in the development of these guidelines, how they have been validated, and how much correspondence they have with the country's social reality or the labor market. It has also studied the competencies implicit in these statements and their demands for students, educators, academic management, and parents.

Curricula. Curricula translate goals and objectives into the activities and educational practices that students will be exposed to learn. Curricula are the road map that educational systems use to guide instruction. Educational research has been used to understand the processes of curriculum development, validation, implementation, and effectiveness. The rationale, content structures, sequencing, and coherence to promote learning and their significance in achieving the goals, objectives, and ideals they pursue have been investigated and evaluated.

Teaching strategies. Curricular activities are accompanied by teaching techniques to provoke learning. Teaching techniques are all the strategies that educators use to educate their students. Understanding teaching strategies and how they promote learning has been one of the emphases of educational research. Research on the effectiveness of educational practices has focused on whether they produce the desired results and whether they contribute to achieving educational goals and objectives. Two lines of educational research fall under the category of the effectiveness of educational practices: (a) **Construction and validation of educational practices.** This line of research consists of studying the construction, validation, and implementation procedures of teaching techniques, study modules, curricula, or programs to establish their scientific basis or validity. Studies of this nature range from preschool education to university education. (b) **Research on the effectiveness of educational practices in real educational scenarios.** This line consists of studying the effectiveness of educational practices in the range of social dynamics that occur in classrooms and schools. This aspect entails delving into the multiple social relationships and environmental interactions that coincide in educational systems: teachers−students, students−students, students with the subject matter and with the school curriculum, and students−parents.

Learning assessment strategies. Assessment for learning is the tool for determining the effectiveness of education. Assessment for learning refers to all those techniques and practices that educators use to assess their students' learning and determine educational goals and objectives. Educational research has been used to study, develop, and validate assessment strategies (objective vs. subjective tests), learning assessment models (diagnostic, summative, and formative), and assessment instruments (standardized achievement and vocational interest tests).

Research on learning. The purpose and product of education is student learning. Learning is examined as an individual student phenomenon and as an institutional product:

Learning as student changes. Learning as a unique phenomenon is investigated from many perspectives and academic disciplines such as biology, social sciences, and education. Research on learning as a particular phenomenon has been to understand what it is and how it occurs. From a biological perspective, learning has been studied by understanding how it happens through the different stages of human growth and development. Learning has been investigated from a psychological point of view, trying to understand the thought processes through meta-cognition, emotions, and motivations for learning. It has also been studied from a sociological perspective or how the individual learns and develops from living in a society and being exposed to its social institutions, such as the home, the community, and the church.

Similarly, learning has been studied from a social anthropological perspective by comparing groups to understand how values and culture affect this phenomenon. The most recent approaches to the study of learning come from neuroscience. The brain is studied as a human organ to understand its functions and its changes due to learning.

The study of learning has generated theories that are used to guide teaching, materials development, and curricula. These theories are also used to develop other research to determine what learning is and how it occurs conclusively. In the field of education, educational researchers have to resort to all these optics to understand, capture, and explain the social complexity that occurs in educational institutions and to be able to explain the changes in students. In academic research, learning is defined in terms of the individual and social changes that occur with students due to the education they receive. Students' behaviors, values, attitudes, habits, expectations, and lifestyles are examined. These aspects are studied in educational scenarios and in the students' families and communities. We seek to understand how education influences student behavior and how the influences of the home, the community, or the church affect student behavior and education. The study of student changes places the educational researcher in front of the complexity and unpredictability of human beings. The analysis of human behavior is complex. The human being is body, mind, spirit, and emotions. The most common educational research approach to the study of student development has been through standardized tests to measure and compare changes in values, attitudes, and skills. These data are used to determine educational effectiveness in achieving goals and objectives related to the development of values, attitudes, and aptitudes.

Learning as an institutional product. Education occurs in institutions that are designed to provoke student learning and to achieve the educational goals set by the country. Research is used to study the "effectiveness" of educational institutions and determine whether they achieve their goals and objectives. Explaining learning as an institutional product implies investigating the "teaching–learning" relationship generated in the educational institution from its policies to the educational practices they establish. Educational institutions constitute the scaffolding that regulates teaching and learning through educational policies, managerial styles of administering institutions, and supervision of teaching. Educational research in this aspect has a large component of institutional research. Many approaches and models have been used to determine learning as an institutional product:

Standardized testing programs to measure knowledge. Standardized testing programs consist of the use of standardized tests to measure student learning in various academic subjects. These tests are used to compare scores among students in the same school (different graduating classes) or to compare students from other schools, school districts, school systems, and even countries. These results are used to determine the achievement of educational goals and objectives and as an indicator of institutional effectiveness in promoting learning (Ponce, 2014c).

Research-evaluation of program/institutional effectiveness. In this approach, the researcher seeks to generate data that will allow him/her to determine the effectiveness of an academic program or educational institution in promoting learning. The objective is to generate an explanation or theory of learning in the context and particularities of the educational institution. To this end, the following institutional components are examined and questioned: (a) the institutional **context** of learning—for example, what educational goals it pursues and what the educational needs of the students are; (b) **input** or an analysis of the means that are necessary to achieve the educational goals established by the institution—for example, an evaluation of what resources are necessary to achieve the established goals about the existing resources available to the institution (teacher as a resource, time needed to achieve the goals, materials, and fiscal resources that must accompany the teaching); (c) **process** or the analysis of how the program or institution allocates the resources necessary for teaching to occur, the effective participation of students, and the use of the institutional resources allocated to achieve the goals; (d) **outputs** or the evidence generated from the process allows determining whether the goals are achieved or the desired results are produced. This type of research-evaluation is prevalent for investigating public school learning (Greene, 2007; Mertens, 2005).

Qualitative research-evaluation. In this approach, the researcher seeks to understand the teaching processes and the institution's functioning from the constituents' point of view. Through field observations and interviews, the researcher generates a theory about the program's functioning or institution to explain its results or effectiveness in promoting learning (Ponce, 2014b).

Institutional learning assessment. The assessment is a four-phase model where data on learning is collected to understand and improve it, if necessary. The process begins by defining the knowledge to be assessed in the institutional mission (Phase I). Phase II collects information about learning as it occurs in classrooms, academic programs, and student support services. Phase III analyzes the data and determines if it resembles the learning objectives or expectations in the institutional mission. Discussion of the data may be by faculty, administration, or researchers. An action or actions to improve learning are formulated and implemented. In Phase IV, data are collected on the effectiveness of the improvement actions, and a determination is made as to whether learning was improved. If learning is enhanced, the assessment cycle is closed (Ponce, 2014c).

Research-evaluation of institutional products. This approach examines the effect of education from an administrative and accountability perspective to determine the effectiveness of educational institutions in producing learning. The researcher seeks to generate an x-ray of the institution's energy. For example, aspects such as how much students learned compared to standards and standardized tests, how many students managed to graduate, what academic averages they achieved, or how many went on to university studies are investigated. This information is used to develop a historical profile of institutional performance and evaluate its effectiveness or compare it with other similar institutions.

Research-evaluation of institutional processes. This approach examines how educational policies, working conditions, teaching resources, and the teaching structure for the development of teaching or learning assessment regulations facilitate, hinder, or affect academic and institutional development effectiveness. This information improves the institution's internal processes and conditions and makes them more effective and efficient.

Research on the influences of educational contexts on the student. Educational research is contextual because it occurs in educational institutions. Educational practices and student learning are studied as individual changes and as institutional products. To understand this, we examine how students or groups of students interact in specific contexts such as classrooms, schools as social institutions, or communities where schools are located as settings for a range of other social interactions. In these contexts, the aim is to understand how educational practices, curricula, or programs affect students.

Educational institutions provide the context and setting for educational research. Educational research must always keep in mind the range of factors that affect the functioning of educational institutions when explaining teaching and learning: the physical environment, the social, cultural, and economic realities of their students, and the political influences that govern educational systems and regulate their functioning (Shavelson and Tone, 2002). For example, the dropout phenomenon is the same in any school, but the reasons for dropouts may be very particular to the social context of the school. One area that can be explored is whether the school is urban or rural. The educational researcher needs to question what similarities and differences may explain learning between secondary school students in an urban school and a rural school. Explaining the particularities of educational contexts is a hallmark of educational research when studying teaching, learning, and changes in students. This poses great challenges when comparing data on teaching, learning, and student change across schools or educational systems and generalizing the same by the historical time of the study, the school's culture, or the student population studied (Ponce, 2014b; Shavelson and Towne, 2002). Several lines of educational research are visible in explaining the effect of educational contexts on the student:

Research on the social dynamics of education. All educational systems develop around a range of social dynamics of where work and learning occur: (a) educators with students, (b) students with students, (c) educators with educators, (d) educators with educational administrators, (e) students with parents, and (f) the educational institution with the community. These relationships are framed by values and political ideals that come in goals/objectives and curricula. These relationships are also prepared by the values and beliefs of its actors. Learning in an educational system is a product of these multiple relationships and not merely teaching educators in the classroom. Understanding how these various relationships shape educational systems and affect teaching and determine to learn is a topic that educational research has addressed. For example, these multiple relationships and social dynamics have been investigated in the specific contexts of the classroom, the school as an educational setting, on school grounds, in the communities where the educational institution is physically located, and in its extracurricular, sports, or social activities. Educational research has also delved into studying the social groups formed in these multiple relationships: sports groups, educational clubs, groups of friends, rivalry, or antagonistic groups. Again, how these groups are formed, what sustains them, and what impact they have on the formation of students is a question that persists in educational research.

Research on the problems of education. Education in the form of the educational system has shown the challenges and problems in its development to produce student learning. Educational research has delved into these recurring problems of education: (a) **Student motivation.** A common problem in all educational systems is students' interest and motivation for their studies. Hearing students mention that they do not like their classes or find them boring is a daily occurrence in many schools. Educational research has studied this problem to understand what it is, what causes it, and what strategies can motivate students to learn. (b) **Learning disabilities.** Some students learn more than others and why some do not learn is another question in the field of education. The study of learning problems is extensive in educational research and a constant challenge due to the large number of social, economic, and political factors that affect education and the realities of students' lives. (c) **Behavior and discipline problems.** Student discipline problems are also visible in educational systems. These problems range from behaviors such as interrupting the teacher while dictating the class, disobeying school rules, to physical aggression to other students or teachers. (d) **Dropout and retention.** Student dropout is the abandonment of school and studies. In some educational systems, dropout is lower than in others. Educational research has studied the reasons for dropping out, the factors that lead to it, the strategies that have been used to prevent it, and the social and economic consequences of this phenomenon. Despite this, the problem of dropout remains a constant in many educational systems.

Research on educational management. Educational management is responsible for the operation of schools. This includes managing finances, administrative and academic affairs, the organizational climate, and creating working conditions that enable teachers to work and students to learn. Educational research has examined aspects of educational management such as the following:

Research on management practices. Different managerial structures organize public and private educational systems. These administrative structures regulate the dynamics among their members and how they are managed and supervised. There are different administrative and supervisory positions in the various educational systems. Educational research has been used to study the multiple aspects of educational management and its impact on the effectiveness of education: the organizational structure of the various educational systems, management styles, and supervisory and leadership styles.

Research on educational policies. Educational policies constitute the norms that institutions use to regulate their functioning. Policies are the "rules of the game" that define the expected behavior of the members of the educational system. Policies regulate the functioning of the educational system by establishing the dates, procedures, responsibilities, and duties of personnel in the system's processes. Policies define the vision that the human resources must follow for the educational system's success and regulate all the functions of the educational institution such as teaching−learning, student evaluation, and relations with parents and the community. Since the 1990s, the study of educational policies has gained interest due to their impact on the effectiveness of education.

Research on technology/educational resources for institutional effectiveness. This line includes the physical facilities of the educational institution, equipment, and materials for teaching and administrative tasks. The study of technology to support instruction is gaining interest in this era of information technology. It is also studied how teaching support resources affect student learning and institutional effectiveness.

Research on the effectiveness of management practices. Two lines of research are visible on the administration and management of educational institutions. (a) **Research on managerial effectiveness and efficiency.** This line of research studies the effect of administrative practices on institutional effectiveness and efficiency as reflected in the achievement of goals and objectives, the management of finances, the organizational climate, and the satisfaction of employees, students, and parents with the institution. (b) **Research-program evaluation.** Subjects of study develop education. These subjects are organized by academic programs that determine where student learning begins and where it ends. Curricula vary in the content to be studied, the competencies to be developed, and the evaluation strategies. Program evaluation is another strategy for determining learning and educational effectiveness. Program evaluation is a common practice in the field of education. Program evaluation uses educational research to study and understand the effectiveness of one aspect of a program or all of its components. From a managerial perspective, program evaluation is considered a criterion for determining fiscal efficiency concerning how students perform academically about the program's funding.

Educational Research Strategies

Educational research uses six strategies to investigate education. These strategies are as follows. (a) **Observing live educational events.** It means entering educational settings to directly observe educational phenomena has been a logical and effective strategy of educational research. Through observation, educational researchers collect data on behaviors, emotions, processes, events, dynamics, or the physical and emotional environment they manifest themselves. (b) **Interviewing.** This consists of a guided conversation about the subject being studied to understand it. There are occasions when the researcher observes the behaviors, emotions, processes, events, dynamics, or the physical and emotional environment in which the studied phenomena are manifested. Still, it is not until he talks to the protagonists that he/she can understand them. For example, at first glance, school dropout seems to be the same behavior in any educational institution. However, it is produced by various factors that may vary from one educational institution to another. The interview is a common technique to accompany observation. The interview is also used to study behaviors, emotions, processes, dynamics, or the physical and emotional environment of events that have already occurred. Investigating events *ex post facto* or after they have occurred is a common approach in educational research. Education is alive and occurs daily in schools. Therefore, many of its events, phenomena, and problems can be identified once they manifest themselves. (c) **Experiment.** Educational researchers sometimes intentionally create situations and provoke dynamics to observe the events or behaviors they wish to study to answer their research questions (experiments). In education, most experiments occur in educational settings because of the difficulty of developing them in laboratories and under controlled and idealized conditions, as occurs in the natural sciences. The experiment is commonly used to determine the effectiveness of new teaching techniques, curricula, or programs. (d) **Surveying.** Education involves students, teachers, principals, parents, and the community where schools are physically located. The survey consists of collecting data from many people or groups of people on the same topic. For example, what is your opinion about the 7:30 am school start time policy? By collecting data from many people, educational researchers can identify people's views, attitudes, and values in a school, district, or educational region on the topic being studied. Surveys use questionnaires or structured interviews (oral questionnaires) as a data collection technique. In these surveys, study participants answer the same questions, making it easier for researchers to identify the positions that exist on the research topic. (e) **Studying particular cases.** Extraordinary events occur in education that stand out from the norm and capture their constituents' attention because of their positive or negative consequences. For example, a student perseveres in his or her studies despite the adversity of life, a teacher who achieves extraordinary results with his or her students, or a school that gets its students to graduate and pursue university studies. The same example can be applied in reverse. It is called a case study when the focus of the research is on that one student who perseveres in his or her studies, that one teacher who achieves extraordinary results, or that one school with admirable achievements. The case study is oriented to identify those particularities of the "case" being studied to improve other students, teachers, or schools. (f) **Studying history.** It consists of studying educational, social, economic, or political events of the past to understand their implications for the present. Historical studies in education seek to understand the origin, development, facts, and interpretations of events, programs, educational practices, or policies and how they influence contemporary education. The historical study resorts to analyzing documents and interviews with people related to those events of the past. If they are alive, they can help the researcher know those events that they lived or knew. Until the 1990s, historical studies were a common approach in educational research (Pring, 2000).

Controversies

Controversies with educational research arise from its institutionalization. All educational research originates from the interest of someone in its institutional structure (e.g., politicians, administrators, or professional groups). Someone conducts the study (an academic, a consultant, or a specialized group). A product is generated (e.g., a research report) that may please some and displease others because of the implications it carries (Carr and Kemmis, 1985). Throughout its history, controversies with educational research can be classified into three categories:

> **Administrative controversies.** Although educational research is justified about its contribution to education development, any educational reform always entails changing or modifying its academic or administrative structures (Carr and Kemmis, 1985). The expectation has been that educational research will generate data that will be translated into principles that guide teaching practices and educational policies that

will allow the scientific management of educational systems (Vinovskis, 2009: Condliffe and Shulman, 1999: Carr and Kemmis, 1985). Throughout its history, there are documented episodes where politicians, academics, or administrators have expressed dissatisfaction with the quality of education and the effectiveness of educational research to reform public education or solve its problems. These criticisms have questioned the quality, usefulness, or scope of educational research (Barnhouse Walters, Lareau, and Ranis, 2009). Actions to address these criticisms tend to restructure, eliminate, or create government offices to regulate and oversee educational research. These restructurings are accompanied by laws that regulate their practices and methodologies (Vinovskis, 2009). This topic is discussed in depth in Chapter 2, "The Problems of Public Education and the Need for Research to Improve It."

Methodological controversies. Scientific research is developed around methods and procedures. Questions about the effectiveness, quality, utility, and scope of educational research have generated debates with the research methods that have been employed in academic research. Educational research uses three models at the time of writing, the quantitative research method, the qualitative research method, and the mixed research method. Methodological controversies have centered on the superiority and relevance of one way to investigate education and its problems (Ponce, 2016). Three issues are recurrent about educational research methodologies; the validity of the data they generate, the generalizability of this information to transform education, and the relevance of current research methods to capture the complexity and dynamism of education (Ponce, Pagán-Maldonado, & Gómez Galán, 2018). This topic is discussed throughout this book in Chapter 3 (Philosophy-based research: The entry of quantitative and qualitative methods into educational research), Chapter 4 (The quantitative and qualitative paradigm debate: Alternate Visions of How to Research Education), Chapter 5 (The Emergence of Mixed Methods Educational Research), Chapter 6 (Science-Based Research to Provoke Evidence-Based Educational Practice: Political Consensus on What Educational Research Should Look Like), Chapter 7 (Education as an Object of Scientific Inquiry), and Chapter 8 (From Research Applied to Practice to Research Embedded in Practice).

Ethical controversies. The institutional structure of educational research is made up of many actors with diverse interests. Inevitably, research will involve the interests of those seeking scientific analysis to solve the problems they address. Ethical and political controversies with educational research can arise to the extent that a study supports the interests of one group and disfavors others. For example, a study indicates that an administrative measure will benefit education, and your organization of educators believes that the study's findings are adverse to teachers. This group may argue that the research was the product of a political assignment to advance the education bill. The ethics of the study is an evaluation criterion in the institutional structures that regulate educational research. With the implementation of institutional arrangements to control ethics in educational research, many controversies have arisen between researchers and the officials of these bodies. The conflict emanates from the demands for methodological modifications or research procedures that these agencies require of researchers to authorize their studies. This issue and its methodological implications are discussed in Chapter 9, "Educational Research in an Era of Ethics in Scientific Research."

2

The Problems of Public Education: The Need to Research Education to Improve It

Omar A. Ponce

Educational research emerges at the beginning of the 20th century from the interest in solving the problems of public education (Ponce, 2016; Ponce, Pagán-Maldonado and Gómez Galán, 2017 and 18). As we know it today, public education emerges as an essential strategy in the social and economic development plans of many countries around the world in the late 19th and early 20th centuries. Education has a direct impact on health, nutrition, employment, and civic life. It is considered the driver of all 21st-century goals because it equips people with the knowledge and skills to break the cycle of poverty and build a better future for themselves (Bokova, 2010). In the form of a social institution called school/university, education generates problems that impact the aspiration to prepare the new generations of citizens and employees of a country. This chapter examines public education issues as the first major event from which the interest in researching education emerges.

Education as an Instrument of Social and Economic Development

The interest in using education as an instrument of social and economic development can be traced back to the 18th century through concepts such as "popular education," "education for all," "education for the poor," and "public education." For example, the concept of "popular education" took root in the 19th century, the context and emergence of National Education Systems in various countries. In the 20th century, the concept of "popular education" became related to the idea of "education for all" or the "education of the underprivileged" or oppressed groups, such as the poor. "Education for all" is based on the ideal of an ideal citizen. For example, in France and the United States of America, this ideal is operationalized and embodied in the Declaration of Independence document of 1776: preservation of life, respect for liberty, and the pursuit of happiness. The problem was that not all citizens had the same opportunity to pursue happiness, as some were not considered citizens. For example, in 18th-century French society, social class differences existed between privileged citizens and the populace. The problems generated by certain social groups were considered a threat to the Republic and had to be addressed. Addressing them implied educating people to eradicate these problems. By 1796, the French Society issued the principles of what later became the country's first educational policy. It delineated the relationship between the state, citizens, and education. It formulated the following principles and objectives of the country's elementary education system, as did Holland: (a) every person must fulfill his responsibilities as a citizen—this includes knowing the language of his country; (b) to be able to communicate his thoughts in writing; (c) he is proficient in arithmetic to attend to his affairs; (d) to be aware of his responsibilities to divinity, and himself and others; (e) to be knowledgeable of the laws of his country so that he can comply with them as a citizen. The state was to have an educational system with its metrics for improvement. These metrics consisted of (a) a school to train teachers, (b) a system of inspection of teachers, (c) an educational system for the poor financed by state taxes, (d) a good salary for teachers, (e) a compulsory minimum education, and (f) a basic and classical curriculum for the initial grades. In 1796, a year after the government takes control of the country, New Zealand presents the first draft of its constitution, which provides for moral and civil education, the arts, sciences, and all skills necessary for the access and development of all citizens. The idea of education accessible to all was the beginning of the concept of "popular education." In 1800, all Dutch schools accessible to all were called "public schools." The concept of "public school" is still current at the writing of this paper in many countries, not only to describe a school for the poor but also for "everyone in society." Although there is no universal definition of what public education means, the common usage seems to be an education financially supported by public funds (Braster, 2011).

The concept of "education for all" came to be seen in the 19th and the early 20th centuries as "education for the poor." From the 19th century, "education for all" was understood as a means of eradicating some social issues, such as poverty and immorality, among the poor. This required the teaching of valuable skills and sciences to promote virtues. To educate, it was necessary to establish an education system, improve teaching methods, train teachers, provide lectures to educate people on many subjects, build libraries, and found a bank for savings. Education for all is the basis of contemporary public education in many countries. The concept of "education for all" involves several activities that include the poor sectors of society. In many countries, some people established models of education for poor children; for example, Maria Montessori in Italy, Paulo Freire in Brazil, or José de Calasaz in Spain. The concept of "education for all" has been confused at times in history with education for the poor. For example, in the early 20th century, education for all was understood as a form of popular education where people learned skills for employment. From this vision, universities emerged to provide training to help graduates achieve social mobility. Between 1848 and 1939, many educational movements emerged in Europe and inclined to the idea of education for all or popular education (Braster, 2011). The same occurred in the United States of America with formal education or typical school (Johanningmeir and Richardson, 2008). Formal education was perceived as the new church of the state to bring about an everyday citizen (Goldstein, 2015), and provisions were made for this to occur. Brodisky (1999) identifies the following ten events as the most significant in the development of public education in the United States of America: (a) the GI Bill of Right or the right of every veteran to free education, (b) racial desegregation in schools where everyone could study regardless of race, creed, or skin color, (c) the Education for Every Child with Disabilities Act, (d) federal financial aid, (e) the use of standardized tests to measure learning, (f) the teaching of reading, (g) the use of computers in classrooms, (h) the development of teachers' guilds, and (i) the establishment of public high schools.

From the 20th century to the present, the interest of having education for all has been to combat poverty and other social ills that emanate from this reality. For example, poverty is a serious problem affecting thousands of children in Africa, Chile, Costa Rica, Cuba, Uruguay, Bolivia, El Salvador, Guatemala, Honduras, Nicaragua, Brazil, and parts of Mexico (Coben and Llorente, 2003). In these countries, education is the key to peace, development, and social stability. Education ensures citizen participation in the economic processes of the nations. This is why education must consider all sectors of the population, including adults. Many countries have adopted UNESCO's goals on education as an instrument to eradicate poverty and have established precise metrics to show progress in improving the quality of life of children. Many countries have also legislated to make education a fundamental right and free of cost for their citizens (Matsuura, 2007).

The Problems of Public Education

To implement the ideal of education for all, it was necessary to legislate public policies and make education a legal requirement for all children. It also entailed hiring and training teachers, establishing educational centers that were later called schools, establishing educational policies to regulate their operation, setting up administrative systems to manage schools, and hiring school principals. Translating the ideal of education for all into an educational system that responds to society's social and economic problems has proven to be a complex and fiscally onerous challenge. Although there are many challenges facing education as a social institution, five controversies stand out in this regard: (a) the quality of education, (b) problems of student retention, (c) low graduation rates, (d) conflicts in philosophical visions of how education should be, and (e) its financing (Bokova, 2010). We call these the structural problems of education.

The Problem of Educational Quality

Educational quality is the term used to describe the expectation of education. Thus, academic quality can never be understood as a stable, given, and delimited quality. Quality presupposes an added value, a competitive strategy. Quality presupposes the result of the work and effort of all and allows a continuous improvement of the product or service. It implies employee commitment, ongoing in-service training, and teamwork. The concept of educational quality is related to neoliberal economic policies that sought social development and economic progress. This sense emerges from the financial crises of the 1970s and 1980s and is influenced by conservative political groups. These policies began to take shape in 1940 with the idea of a state that would control access to financial markets. Consequently, the concept of new liberalism that would exercise control over public spending, manage the economy, and improve unemployment constituted a political vision of social and economic development. This development was possible in the competitiveness and opportunities seen in the global market where the struggle is to assume financial leadership (Colella and Diaz, 2015).

The concept of educational quality that emerges with the argument of a global economy is quality as a lack. "Academic quality as a lack" means that quality is constructed in the presence of a deficiency or a void. Quality is not whether education meets educational statements (educational effectiveness) but how it responds to what is expected or desired or "as it should be."

Quality is assessed in terms of a gap based on the state of education affairs at present. Improving the quality of education is always working toward the desired ideal state and not the state of the past. This justifies intervention and continuous improvement actions. Improving quality presupposes a deficiency or loss in the past that must now be corrected. This presupposes an expansion of the system and a historical change toward the future. Change is always upward and forward. Change is always part of development and is inevitable to respond to changes in society. From this perspective, the two significant educational changes with the neoliberal movement, from the 1980s to the present, are educational infrastructure and curricular content. The emphasis has been on schools having the classrooms, equipment, and technology for advanced education in terms of infrastructure. In terms of the content to be taught, the emphasis has been on courses linked to the world of work, courses aimed at training students to build a progressive society, and greater exposure to science and mathematics as the fundamental knowledge for a technological society. The quality of education as a problem is conveyed under the concepts of globalization and internationalization. This presupposes that there is only one economic and social order globally because we now live in an information and technological society. Reforms in educational systems are accompanied by changes in teacher preparation programs to meet the desired quality expectation. The emphasis on quality shifts from the process to the product to have a graduate with the necessary competencies. From this perspective, educational reforms have been statements of the changes to be made in schools to improve the quality of education. Educational reforms focus on transforming systems to alter the human resource in correspondence with the financial world.

Providing quality education for all, or where students learn and develop according to plan, entails the presence and conjugation of several structural and management elements of that structure (Ponce, Pagán-Maldonado and Gómez Galán, 2017 and 2018). In structural terms, the following are required: (a) prepared or trained teachers; (b) a reasonable number of students per teacher; (c) working with parents in those communities of poverty so that they can help their children; (d) a safe and healthy learning environment; (e) restrooms for students; (f) sufficient educational materials; (g) a student-centered curriculum; (h) sufficient teaching–learning hours (800 hours per year); (i) teaching in the child's vernacular; (j) a transparent and accountable school (Matsuura, 2017); (k) poor nutrition continues to affect learning in many countries, making it necessary to have school feeding programs (Kless, 2017); (l) in administrative terms, there is a need for good educational leaders and administrators, development vision, clear work goals, and coherent educational policies (Ponce, Pagán-Maldonado, & Gómez Galán, 2017).

In some countries, such as the United States of America, public education in its beginnings faced problems of racial segregation, religious diversity, nationality, and social classes due to internal and external emigration. Public schools received sick and hungry children due to the lack of medical and food services in which they lived. They received children with learning problems and physical and mental disabilities that brought to light the multiple complementary services that should accompany this public education to facilitate the socioeconomic aspirations of the country. Many of these services evolved into the food and special education programs visible in some countries. These student needs have required more trained teachers, profound educational reforms, legislation to implement them, and the integration of parents and the community to achieve their effective implementation. Promoting academic quality that facilitates student learning in public schools is and will continue to be a challenge.

Since the 1980s, education has been changing in response to the need for improvement to enable the country to have more skilled employees for a global economy. Many countries have reformed their education systems to position themselves for the new global economy. Contemporary education demands much more complex thinking skills from K-12 students (March, 2015; Phillips, 2015). Many educational reforms have been ineffective, not because they lack scientific foundations but because they do not address the problem of poverty and social inequity that limits access to real learning opportunities early in life (Phillips, 2015; Shepard, 2015). Others understand that many educational reforms are fragmented because they do not consider all components of educational systems (Smith, 2015). In the instrumentalist view of contemporary education, to produce employees for a global economy, values are not compatible with the very nature of education, as is the development of the student (Phillips, 2015). Contemporary education seems more like training than education (Pring, 2007). In this instrumentalist view, teachers are penalized for not reproducing the ideal of job training in the form of quality standards (Phillips, 2015). There is a belief that the school's expectation is too high in pretending that this institution alone will change the problems of inequity and poverty in the countries. It seems that there is still no clear vision of that school and that education is desired (Smith, 2015). Others argue that education needs to be rescued from administrators and politicians, and the vision of education for economics and employment, and returned to the idea of education to develop the learner (Alberts, 2015; March, 2015; Phillips, 2015). The world's problems are too complex. Contemporary education does not produce the leadership, innovation, or knowledge needed to solve the current issues experienced by humanity (Gee, 2015).

It is argued that contemporary education for all needs better learning assessment systems to understand student learning and attitudes. It needs educational research with the ability to generalize to the real world. The relationship and interaction between teachers and students need to be rethought and restructured because it is not effective the way it occurs (Kless, 2018). Motivation problems among students, discipline problems, and learning problems are visible in educational systems worldwide (Ponce, 2016; Ponce, Pagán-Maldonado, & Gómez Galán, 2017 & 2018). Contemporary education seems to have been reduced to information literacy and numeracy. Education needs to embrace a broad concept of what is happening in society (Kless, 2018). Schools need to be more inclusive to accommodate children with disabilities. Teaching needs to be a better-paid profession with working conditions that are more sensitive to the educator's work (Bokova, 2010).

Since the middle of the 20th century, the quality of education worldwide has been measured based on students' scores on standardized tests. Explaining a student's scores on a standardized test has implied a profound reflection on the phenomenon of education and the reality and context in which it occurs. Educational systems pursue the common ideal of developing students' capabilities. Not all children who arrive at public schools are the same, nor do they learn the same. Education systems do not respond to the diversity of students and the learning styles they receive. This makes it evident that there is a difference between the ideal of "equal access to an education" and "the provision of educational service." According to Marks (2002), two problems arise from this educational structure: the problem of student retention and the problem with graduation rates.

The Problem of Student Retention

Education and social equity policies in many countries have facilitated free public education for millions of children. Many countries seek quality primary education, promote gender equity, and provide early childhood education and lifelong learning as strategies to deal with combat poverty and social inequality. To achieve access to education, many countries have removed economic barriers so that all children have access to free primary education (Matsuura, 2007). However, getting students to graduate represents a severe problem: (a) 72 million adolescents do not complete secondary school, and this translates to 759 million adults with few job skills, unemployed, or marginalized (Bokova, 2010), and (b) one (1) billion children around the world do not attend school when they should (Kless, 2018). Student retention is one of the most researched and unresearched topics at all educational levels.

The Problem with Low Graduation Rates

Once children begin their formal education, education systems face the phenomenon of the transition from school to the world of employment and accounting for the number of children who complete their studies. Education is an essential ingredient for a good life and the job they seek for many young people. The post-secondary or college career choice they pursue is defined by their academic abilities, personal interests, the opinions of their parents, and the interests of their friends. Today's youth are faced with many career choices, and many fear making the wrong career selection. Research indicates that school plays a vital role in helping young people make career choices. Studies suggest that children from families with college-educated parents tend to complete the public education process. Works also indicate that the longer a student spends in the educational process, the more likely they are to drop out because their social realities change, creating new challenges and difficulties. An objective in many educational systems is to ensure a logical and reasonable transition from school to the world of work. Educational policies seek to ensure this transition. The transition process has been thought of in the linear logic that the student makes a study decision and pursues it from school to the world of work. In many schools and universities, this is reduced to academic indices and educational preferences. Study choices are rooted in cultural factors (Pless, 2014). By 2030, it is estimated that only 69% of students entering high schools will graduate, only 84% will complete middle school, and 70% will complete elementary school (Kless, 2018). It has been proposed as an educational goal to transition from school to the world of work faster to respond to present social realities and thus optimize existing resources. Despite this, getting students to complete their free public education is another serious problem facing the world's education systems (Pless, 2014).

The Problem of Education Financing

The government fully funds education for all or public education. There is a position that public education should be viewed as an investment, not an expense, because a child, regardless of race, nationality, or economic status, has the opportunity to learn and develop. The fact is that sustaining a free public education system is onerous because it entails the following: (a) acquiring buildings or constructing them to convert them into schools, (b) maintaining these buildings in good condition for their uses, (c) hiring teachers and educational administrators and covering their salaries, salary increases, and job fringe benefits, (d) paying monthly water and electricity bills, (e) purchasing new educational equipment and replacing old ones, (d) paying the salaries of support employees in the schools such as social workers, psychologists, clerical staff, and food programs. In some countries, the quality of public education is linked to the refusal of incumbent governments to invest the money needed to operate an optimal public education system. For example, in the United States, Mathis (2003) argues that after analyzing the implementation of the No Child Left Behind Act in 10 U.S. states, the federal government was demanding too much compared to the money allocated.

In the United States, the attack on public education by economic interest groups has led many to believe that teacher quality is not in their college degrees. Having tenured teachers is the problem with education in charter schools. And vouchers are the alternatives to improving education because parents can take their children to their choice and schools without teachers' unions. Private instruction is better managed than government schools. The result is public money flowing to private entrepreneurs (Kilty, 2015). Education for all or public education has a cost, and more economic injections are needed to strengthen its financing and operation (Bokova, 2010).

The Philosophical and Visionary Problems of Education

According to Rogers (2012), education is conceived as training for civic citizenship in a society, such as education for human survival, employment or vocation, adjustment to life in community, self-expression, social reform, or spiritual development. In practice, one approach always dominates another. Throughout history, different leaders have viewed education differently; in the 19th century, education was understood to equalize opportunities in society by training conscientious citizens. In the social Darwinist view, education was to prepare the child for life in industrial society. Rousseau, in his romantic vision of progressivism, saw education as a form of self-expression. In the case of American education, there has never been a consensus on what public education is. Public education means different things in different countries. In the U.S., it means education that is funded, implemented, and administered by the government. This education is supposed to be ideologically neutral. In the case of the United States of America, 150 years ago, the American elite encountered waves of Catholics, immigrants, and the poor, which led them to seek alternatives to standardize this diversity and generate a single person from all these groups. One way to achieve that ideal was to create a typical school to create the nation's citizens. So common has this vision of education become that it is difficult to conceive it any other way. In practice, public education in the United States of America reflects progressive ideals as a philosophy of education that the American people have historically rejected. It is no wonder that public education is a subject of constant controversy. Although public education is debated in other countries as well, the view is that public education is a pluralistic education or education for all groups with diverse beliefs but serving society. In the United States, secularism has been confused with neutral in matters of education. The secular view of American education has dominated public education since the mid-20th century. This has caused religious groups to develop parallel educational systems.

Other democracies in the West have developed standards and revenue sources to support diverse religious and secular groups with different philosophies and visions of education. For example, the Netherland government supports over 30 different types of schools. In England, 60% of children attend publicly funded Jewish schools. In Italy, 25% of schools are government-funded schools with diverse educational philosophies. In Israel, schools are religious and secular, where teaching is Hebrew or Arabic and where between 55% and 75% of schools are government-funded. Educational diversity is also seen in countries such as Australia and Switzerland. The common element that integrates all these schools is a national/regional curriculum and assessment commitment. Children from different classrooms are exposed to similar civic and academic experiences and where common, rather than prescriptive, themes are established. Pluralism in education means teaching a variety of subjects in society, religious and secular, but respecting the particularity of each group in society. It is not a slogan to privatize education but to develop the new generations of a multicultural society. A pluralistic education will not solve all the problems of American society, but studies show that it produces a more complete civic and academic education for living in society. No form of education is neutral, but it can be honest in recognizing the values on which it is based. Two controversies emanate from how child development and teacher involvement in this regard have been understood:

> **The debate between traditionalist vs. progressive philosophies**. Traditionalist philosophy sees the child as a vessel of knowledge that opens the child's senses and doors to social opportunities. The curriculum in this view tends to be linear, sequential, and chronological with high subject standards. The **progressive** philosophy sees the child as a creator of knowledge and not as a recipient. They favor curricula that allow for the social development and creativity of children. They believe that more open and flexible classrooms foster students' critical thinking, which traditional classrooms do not.

> **The debate between the religious and the secular in schools**. Secularists see the child as an autonomous being capable of thinking and not constraining religious values. The role of the educator is to take care of the neutrality of the process and educate the child emotionally and intellectually without imposing doctrines. The religious see the child as a spiritual, emotional, and intellectual being, and if the spiritual part of education is ignored, it borders on a form of indoctrination.

Since 1920, the progressive philosophy has dominated university colleges of education as well as public education. The controversy with teaching by standards comes because legislators demand more specific teaching and learning assessments from educators. For some, the educational uniformity that is attempted to be imposed is a form of indoctrination, and for others, plurality makes accountability impossible. It is recognized that public education is imperfect. Jurisprudence establishes that parents may enroll their children in the school of their choice. It also confirms that curriculum content belongs to school districts and that parental authority over their children's education is diminished if they consent to enroll in the school. For years, teachers' groups have exerted influence over the content and affairs of public education to the point that they have influenced its effectiveness. Some believe that countering this issue will involve changing the structure of public education. Educational innovations such as charter schools, educational vouchers, and distance education promote diversity in American public education. These innovations challenge the traditional structure of government-transferred public education to government-funded public education. This changes the philosophy and politics of education to a social, civic education that is not government-delivered.

Lack of Evaluation of Educational Policies

Educational policies shape educational systems because they establish how, when, where, and by whom the tasks that need to be done to achieve academic success will be done. All educational systems have educational policies that delineate how teaching, assessment of learning, and relationships among constituents should be. Historically, educational policies emanate from school administration and are imposed on educators, students, and their parents to regulate the operation of schools. Policies tend to be ideological statements developed in light of what the committees responsible for their development understand teaching or assessment should occur. In the development of policies, literature is also researched to support the regulations in available scientific studies or the successful practices of other educational systems in the world. Since the 2000s, interest has been generated. It is recognized as a need to evaluate the scope of educational policies and their effects on the development of quality education (Ponce, Pagán-Maldonado, and Gómez Galán, 2017a and b). For example, Ni and Jia-Li (2015) illustrate how educational policies intended for urban areas in China's public schools turn out to be discriminatory when applied to public schools in rural areas. Evaluating the scope of policies or establishing science-based education policies is essential to improving education (Ponce, Pagán-Maldonado, and Gómez Galán, 2017a and b).

Methodological Evolution

The emergence of public education and its problems had several implications for educational research that are still valid in the 21st century. (a) The conditions were created for the emergence of the interest in scientifically researching education. From this interest and this need, the term educational research was born to refer to scientific research on educational issues (Ponce, Pagán Maldonado, & Gómez Galán, 2017 & 2018). (b) The need for scholarly research that improves education and the responsibility to inform the country and academic administrators to develop better educational practices and policies is established (Johanningmeir and Richardson, 2008). This created a high public and political expectation with the benefits of scientific research in education (Ponce, Pagán Maldonado, & Gómez Galán, 2017 & 2018). (c) It forced educational researchers at that time to identify the model for researching education.

3

Philosophy-Based Research: The Entry of Quantitative and Qualitative Models into Educational Research

Nellie Pagán-Maldonado

The problems of public education aroused interest in universities to investigate education to improve it scientifically. Operationalizing this interest confronted researchers and academics of the time with the challenge of identifying the model to be used to investigate education scientifically. At the beginning of the 20th century, the dominant model of scientific research in universities was the experimental method of the natural sciences. Also, there were research practices with a different name in the social sciences, particularly to their disciplines, but not all considered scientific methods; For example, ethnography in anthropology, historiography in history, and philosophical analysis in philosophy. It is clear from history that scholars in early Colleges of Education in universities did not develop a particular research model for education but initially adopted the experimental model of the natural sciences (Condliffe, 2000; Johannigmeier and Richardson, 2008; Ponce, 2016). This chapter explains the inception, development, and challenges of initiating educational research with models imported from other academic disciplines to scientific research in education.

The Search for an Educational Research Model

The origin of educational research is linked to the emergence of public education systems worldwide during the industrial revolution. With the industrial revolution, many social phenomena and problems occurred in industrial cities. For example, in some countries, emigration from the countryside to industrial cities or from one country to another produced overcrowding in the new metropolis and social segregation between nationalities, races, and social levels. Problems of criminality and healthy coexistence emerge. Many of the social issues during the industrial revolution were understood as educational problems of population sectors that did not know how to live within the new socioeconomic realities. The interest in educating everyone to improve the quality of life and economic productivity of the country led to the establishment of public schools. The establishment of public schools implied creating educational systems to manage and supervise the schools that were established throughout many countries. The most common administrative model for managing public schools was that of educational districts and regions. This model responded in its logic to a school organization based on the geographic location of schools. The establishment of a public education system also entailed hiring school administrators to supervise schools and school districts. With the creation of communities and educational regions came the appointment of school administrators and the phenomenon of public education oversight. Dissatisfaction with student academic performance was the primary concern of many school principals in many general education systems. This culminated in debates about the capacity of many teachers and the effectiveness of professional training programs. Student learning and the effectiveness of teachers in promoting it emerged as the big issue in public education. In one way or another, facilitating student learning became a topic of discussion and public policy that needed to be addressed because it affected the socioeconomic development of countries (Condliffe and Shulman, 1999; Condliffe, 2000; Johanningmeir and Richardson, 2008; Ponce, 2016; Ponce, Pagán-Maldonado, and Gómez Galán, 2017 and 2018; Gómez Galán, 2017).

In the case of the United States, the problems of public education caused some universities to approach public education in three ways: (a) to help colleges and teacher academies improve their training programs, (b) to train school administrators, and (c) to use schools as research laboratories. The university, as a scholarly body, was the best letter of acceptance. Research on education and the effectiveness of public schools in promoting learning become subjects of university studies (Walters, 2009; Johanningmeir and Richardson, 2008; Condliffe, 2000). Between 1855 and 1900, many prestigious universities in the United States establish their teacher preparation and school administration programs: Harvard, Stanford, Columbia, and New York University (NYU) were the first universities in the United States to establish Colleges of Education for the study of teaching. In these decades, university discourse and debate about public education produced the following changes in the nation: (a) a refocusing of the public school curriculum to continue university studies—this made associate degrees and baccalaureate degrees the minimum requirement for many professional jobs; (b) federal funds are established to investigate the problems of public education; (c) the first professional journals in education emerge. Harvard University was the first university institution to publish a professional journal in education, *Educational Review*. Professors were hired to teach university courses on education. These professors came from academic disciplines such as philosophy, psychology, and the humanities (literature, art, music, and languages). Education as a discipline of university studies and as a profession began without a defined foundation and knowledge to practice dispersed between the natural and social sciences. The new university professors of education received the task of working with teachers' training and investigating and solving the problems of public education. The first textbooks published in education emerged (Condliffe, 2000). Each of these teachers brought to the field of education the knowledge and research methods of their disciplines of study. Educational research begins without research methods. The first line of educational research, and controversial, was learning: what it is and how it happens. The earliest work came from philosophy by the end of the 19th century. These works produced practical explanations of how learning occurred. The goal was for the classroom teacher to have the tools to do his or her job effectively. This creates the first division in education: the teacher who practices in the classroom with children and the university academic who researches the problems of education.

Between 1860 and 1900, the range of approaches to the study of public education problems was evident. The first researchers of educational issues came from John Hopkins University. These researchers were physicians by training. They approached the study of education from a hygienist and child health perspective. Their approach used the physician–patient model of medicine to examine what the teacher must know to educate his students. Psychologists used laboratory experiments to study students' learning and mental processes. Sociologists developed field studies to understand the functioning of schools. Philosophers and humanists turned to society's historical and philosophical study to propose statements to guide classroom practice. The term educational research emerges to describe the range of undertaken activities to "investigate" public education (Johanningmeir and Richardson, 2008).

By the 1900s, psychology, especially behavioral psychology, as an "objective science" that studies mental functions and their structures, enjoyed greater acceptance than philosophical postulates when explaining the problems of learning. Laboratory experiments and quantification marked the scientific study of education in child development, human movement, hygiene, and learning (Walters, 2009; Condliffe, 2000). Educational research was a science of testing and statistical measurements on theories and attributes of learning, which resembled research in psychology (Walters, 2009). At the beginning of the 20th century, the problems of public education became acute and reached social criticism. Educators, philosophers, humanists, and psychologists became targets of criticism for not solving public education issues. Psychologists gave up their participation in the development of public education, arguing that psychology was a science and pedagogy was an art. Famed schools of psychology at universities, such as Harvard, which once struggled with their presence in the field of education, sought by all means not to be associated with public education and its problems (Condliffe, 2000). Criticism of the effectiveness of educational research and its relationship to the quality of public education emerges for the first time (Johanningmeir and Richardson, 2008).

The creation of Colleges of Education at major research universities in the United States of America forced education professors to conform to the scientific culture of academia. The first professors of education found it necessary to justify education as a scientific discipline of university studies. Between 1900 and 1920, the first central question was whether education should be investigated as a natural phenomenon or a social–cultural phenomenon. Education is a social construct or a creation of the human mind. As an instrument of social reform, it aspired to develop the individual and their social groups to achieve a better society. Public education originates as a social institution to impact social, economic, and political development ideals. Johanningmeir and Richarson (2008) call these interests the "national agenda" that has guided educational reforms throughout its history: the national defense of the country, the promotion of civil rights and the acculturation of immigrants, the elimination of poverty and cultural backwardness, the moral development of students, the rise of health, and the development of a competitive workforce for a global economy. From its origins, the aspirations for public education have been complex to conceptualize, implement, and evaluate.

By the 1920s, it was evident that the term educational research encompassed many activities to study the problems of public education. Four positions are generated and debated about educational research (Johanningmeir and Richardson, 2008): (a) educational research as the study of the functioning of schools to account for and inform classroom teaching and educational administration—this view eventually became the institutional research division seen in the public education systems of many countries; (b) educational research as a mechanism for generating objective data to eliminate the biases and opinions seen in education; (c) educational research as a means of arriving at final and categorical answers that would prescribe teaching and administrative practices in public education; (d) educational research as a mechanism for generating data to inform organizational decision-making about educational needs. In this decade, discrepancies were evident among educational researchers about the scope of research methods and academic disciplines studying education. In the 1920s, there was a clear need to define educational research and to establish whether it should be a quantitative or qualitative science. Educational researchers at that time resorted to identifying the similarities of educational research with research practices recognized in the academic world and did not elaborate on their particularities and differences (Johanningmeir and Richardson, 2008).

From its inception until about the 1960s and 1970s, educational research was considered a multidisciplinary field of study and social science (Koichiro, 2013). The term "multidisciplinary field of study" was used to imply that education was a field where knowledge from both the natural and social sciences was applied. In many European countries, the study of education was conceptualized as a social science. The focus of this science was to study the social and cultural dynamics that occur in educational systems to produce learning (Segovia, 1997). In the universities of the United States of America, education was understood as a particular discipline of study organized in Colleges of Education. The first university professors attached to the Colleges of Education came from the natural sciences and the social sciences disciplines. These brought research models from their academic disciplines and applied them to public education research (Condliffe, 2000; Johannigmeier and Richardson, 2008; Walters, 2009; Ponce, 2016). Much educational research until the 1970s was research in history, psychology, sociology, and philosophy (Condliffe, 2000; Johannigmeier and Richardson, 2008; Walters, 2009; Green, 2010; Koichiro, 2013). Since then, educational research is a science that debates a terrain of knowledge between the natural sciences and humanistic philosophies (Gil Gil Cantero and Reyero, 2014; Ponce, Pagán-Maldonado and Gómez Galán, 2017 and 2018).

The Entry and Development of Quantitative Research in Education

Between 1900 and 1920, many developments occurred that defined education and its research. (a) The departure of psychologists from the field of education allowed philosophy and the humanities to take the lead in educational research. Many philosophers inserted themselves into the polemics of public education, generated theories, validated them in schools through "field experiments" that helped explain education problems, and facilitated strategies to strengthen the "art of teaching." Philosophical postulates emerge to explain the field of education and its social responsibility. (b) Six educational specialties emerge with their lines of research: educational philosophy, educational psychology, educational measurement, educational administration, history of education, and teaching methodologies. (c) An interest emerges in separating the scientific from the moral because of the precise role of values in education. (d) The scientific research methods in the universities are moving away from the scientific principles of Darwin's evolution. This brought educational research closer to quantification, measurement, and the study of cause and effect relationships. In the 1920s, some behavioral psychologists saw the opportunity to test and empirically validate public schools' theories of causal measurement of personality attributes and human behavior. These studies helped to develop the measurement of learning. At this time, the principles of measurement were carried into educational administration. The measurement of the effectiveness of administrative practices and the use of surveys to make decisions helped strengthen the science of educational administration and the measurement movement in education.

In the 1920s, the revolution of educational research models that we know today took place in European universities. The crucial event was the Vienna Circle's displacement of philosophy and the humanities as the only disciplines for knowing the truth and generating knowledge of the social world we live in. Until this time, philosophy and the humanities were the "sciences" that dominated the understanding of the social world. The Vienna Circle was a group of mathematicians, physicists, philosophers, historians, and sociologists at the University of Vienna, who challenged the dominant philosophies of metaphysics, idealism, and phenomenology. Influenced by the research model of the natural sciences, their challenge was to point out how little validity philosophy and the humanities had as "sciences" because they lacked a "method" to verify the authenticity of the claims they made about truth and knowledge of the social world. The Vienna Circle presented the philosophy of social research called empirical logic. This focused on the presence of a "scientific method" to corroborate any assertion of knowledge and the authenticity of truth. Until that time, the search for truth and understanding of the social world was an intellectual exercise in logic and not verifiable. Empirical logic, eventually known as the positivist philosophy of inquiry, is recognized as the most significant movement in social, intellectual development in the 20th century and of the most significant impact on educational research (Paul, 2005; Pring, 2000).

Early educational researchers adopted the experimental research model of the natural sciences. Experimental research was the dominant and most successful model in university scientific culture. The empirical research method in the natural sciences had its roots in the scientific revolution of the 17th century: (a) the investigator identified a cause and effect relationship; (b) he/she constructed a theory or model to help explain the origin and effect of the relationship; (c) design a laboratory experiment to test the causal relationship, theory, or model; (d) he/she performed the investigation and, by mathematical calculations, validated or rejected his/her hypothesis; (e) if he/she validated his/her hypothesis, the finding became new knowledge. Through this model, biology had a great impact on medicine, chemistry had a great impact on agriculture, and physics had a great impact on engineering and technology. The findings of the natural sciences had several attractions for their applied professions: their knowledge was generalizable and made it easier to predict, control, or prevent situations.

In the social sciences, the experimental method was credited with turning psychology into a precise science in the study of human mental processes. The knowledge of mental processes had that tinge of generalization attributed to science (Walters, 2009). Education adopted the experimental method of the natural sciences. This movement caused the following changes that impacted educational research: (a) it refocused educational research from the search for metaphysical truths to the search for scientific, objective, and verifiable knowledge; (b) it redefines educational research from the intellectual exercise of logic used by philosophers and humanists to scientific methods and procedures, as was done in the natural sciences; (c) it provokes the distinction between scientific and non-scientific disciplines in universities, with some undertaking the quest for recognition and status as scientific disciplines— education was one of these disciplines; (d) the quantitative research model emerged, which dominated educational research from approximately 1930 to 1980. With the philosophy of empirical logic, the quantitative research model emerged and became the scientific standard of educational research until the 1980s. It is based on the following principles and assumptions (e.g., Shadish, Cook, & Campbell, 2002; O'Leary, 2004; Balnaves & Caputi, 2001; Charles, 1988; Cohen & Manion, 1980):

The Philosophical Foundation of Quantitative Research

Positivism postulates that human beings know their physical and social worlds through experiences. Knowledge is information about those physical and social worlds. Human behavior is influenced by the physical and social worlds he/she experiences. It postulates that social reality, and its knowledge, as in the natural sciences, is accurate, objective, and external to the human being. It assumes that this reality is regulated by universal laws of causes and effects imposed on the person and that can be measured. Human behavior is the result of this external social reality that is set on the human being. An example of this positioning would be to say that a child who grows up in a hostile environment becomes an aggressive adult. In this example, the environment influences the person's behavior. Positivist philosophy emphasized the study of those external influences of the physical and social educational world, which can be observed, measured, and corroborated through human experience and behavior. For example, group size or a particular teaching technique on student experience or behavior (learning) can be studied and corroborated. Positivist philosophy argued that it was only possible to learn scientifically those things that could be measured. If they could not be measured, they could not be corroborated.

Quantitative Research Methods

Since it is assumed that social reality is real, objective, external to human beings, and imposed on them, human behavior is reactive. The quantitative research model then focused on searching for those cause and effect relationships that will explain human behavior, such as learning, in educational scenarios; for example, how a teaching method affects learning. The quantitative research model was based on the following principles. (a) In the search for absolute truths (knowledge), and the discovery of those universal laws of cause and effect that regulate human behavior. (b) In applying a deductive model, the researcher tests theories, models, or propositions and verifies them through experimentation. When experimentation is not possible, statistical tests of hypotheses are used to establish the validity of the hypothesis or its probability of occurrence through surveys with random samples. (c) The use of standardized and objective methods, procedures, and measuring instruments excludes the researcher's values and prejudices from the study. (d) Emphasis on verification of the data collected. The presence of the scientific method is essential to evaluate the authenticity of the finding and the knowledge generated. (e) Accumulation of observable and verifiable facts as a mechanism to develop knowledge that allows the prevention, control, or improvement of educational events or situations.

From approximately 1950 to 1980, educational research relied exclusively on the quantitative research model to produce knowledge and develop the profession (Paul, 2005). In these decades, academic research was characterized by the following: (a) research into the measurable and the observable, (b) objective research that excluded the subjective, the moral, and the political, and (c) research that produced the correct teaching and learning methods and practices.

Strengths and Criticisms of Quantitative Education Research

For some researchers, the study of causal relationships constitutes the primary strength of quantitative education research. The quantitative research designs used in the study of causal relationships are the experimental design, the quasi-experimental design, the causal comparison design (*ex post facto*), and the correlation design (Ponce and Pagán-Maldonado, 2016b; Ponce, 2016). Causal comparison and correlation designs are standard in education because many variables cannot be controlled at the time of the study. Researchers have to resort to these designs to study them as they manifest themselves in schools (Johnson, 2001; Ponce, 2016; Ponce and Pagán-Maldonado, 2016b). The causal relationship model has focused on identifying possible relationships with correlation studies that are then sought to be validated with experimental studies (Redford, 2006; Galán, Ruiz-Corbella, & Sánchez Melado, 2014; Ponce, Pagán-Maldonado, & Gómez Galán, 2017a & 2018). For some scholars and researchers, designs with random samples, such as the experimental design, are considered to be of greater strength for studying causal relationships in education because they allow controlling the study variables. The random selection of participants reduces biases in the study of groups and minimizes the possibility of rival explanations for observing causalities (O'Connell and Gray, 2011; Shavelson, 2015). The use of statistics facilitates creating mathematical models of these relationships (O'Connell and Gray, 2011), which can then be accurately controlled and measured through experimentation. This helps to minimize errors in interpreting the causal relationships being studied (Shavelson, 2015; Kaplan, 2015).

The major criticism of quantitative experimental and non-experimental designs in education is reductionist and deductive research approaches. This makes them weak and of little scope to capture the complexity of education (Radford, 2006; Feinberg, 2012; Galán, Ruiz-Corbella, & Sánchez Melado, 2014; Rowe & Oltmann, 2016). Experimental methods are underpinned by a positivist and post-positivist philosophy from epistemological and ontological perspectives, where knowledge is understood as an entity separate from people and reality as an objective entity. For example, the wisdom taught by teachers is independent of students. If the student does not learn, then the cause must be in the teaching methods and teachers. This type of interpretation can distort facts and negatively affect education by comparing schools because it does not consider or capture the complexity of schooling (Rowe and Oltmann, 2016) or the implicit values of the situation (Feinberg, 2012).

The Entry and Development of Qualitative Research in Education

Between 1910 and 1930, many industrialized countries continued to experience the consequences of the change in economy and life brought about by the industrial revolution. For example, in the United States, the consensus that existed on the problems of public education disappeared because of several events that occurred: (a) demographic changes due to foreign immigration, (b) the public policy pronouncement entitled the Cardinal Principles of Secondary Education that urged a redefinition of curriculum content in schools to respond to the social and economic needs of the country, and (c) the economic recession and the crisis of the First World War. The study of the school curriculum emerged as the cause and the alternative to the problems of education. The study of the curriculum, its construction, its content, its evaluation and research, and its relationship with society emerges as the focus of public education. In many universities, the study of curriculum and instruction became a specialty that produced the first doctoral degrees awarded in colleges of education at universities. Many universities developed general education curricula to connect secondary education with the professional education being offered. The argument was that many public school graduates were arriving at universities with significant academic lags. This again raised questions about the quality and effectiveness of public education (Johanningmeir and Richardson, 2008; Condliffe, 2000).

Between 1930 and 1950, public education becomes the largest industry in the United States of America. Colleges of Education become entrenched in universities. As the focus of education, the curriculum made it easier to connect separate pieces of this profession: (a) measures of mental attributes developed by psychologists are aligned with specific curriculum activities and teaching strategies developed by educators and philosophers; (b) it connects and aligns the curriculum sequence developed by educators with the development stages developed by psychologists; (c) attributes of IQ and child development levels, developed by psychologists, are connected with the curricular levels and their criteria for measuring learning created by educators. This brought new variables into the learning equation: what role do interpersonal relationships play in learning, what influence does the student's culture have on learning, and how do the learner's social class and mobility occur due to education. This set the stage for anthropologists to become interested in educational research (Johanningmeir and Richardson, 2008; Condliffe, 2000).

With the academic study and scientific research of the curriculum in the 1950s, partisan politics emerged as a determining factor in the development of public education. The importance of education in the development of youth and the nation's welfare captured the interest of American philanthropists who invested in the development of education through funds or the creation of entities. The Rockefeller Foundation and the Ford Foundation have helped many disadvantaged children to achieve university studies and so did the Carnegie Foundation, which funded educational development projects that are still present today, such as the Educational Testing Service, which administers psychometric tests to determine placement in academic studies at all levels. Since the 1950s, partisan politics in American education is overt in the contest for federal funds. The curriculum and its content represent the battleground where the proponents of a curriculum centered on disciplinary content that contributes to the country's economic development versus a curriculum that prepares the citizen for life and employment and contributes to the social development of the nation polarize. The first group demands excellence in education through high-quality standards, and the second requires equity and access to education. The first group focuses on investing funds for education in science, mathematics, research, and economic development of the nation. The second group advocates the social teaching of the masses that contribute to solving the social problems of the country (language, fine arts, and physical education). The fact here is the limited participation of educators in decision-making on education development at all levels. Actors such as the National Science Foundation and the Federal Department of Education have been influencing the criteria, curricular content, and research type to be conducted at all educational levels. Since 1980, the influence of an education that focused on economic development has generated national programs of measurement, evaluation, and assessment of learning for public schools and universities and national educational research projects based on statistical measurements and accountability. These projects generated controversy about learning because of their questionable results. This reemerged the concern with the inadequate competencies of public school graduates in national tests and the debate with the quality of education and educational research. By 1940, parallel to these events, some anthropologists began to investigate the social phenomena of education with their field research methods. Entering schools to observe and interview to understand education in the various educational settings was a departure from quantitative research methods. These methods eventually became known as qualitative research.

Between 1950 and 1970, qualitative research becomes evident in educational research. Two reasons seem to explain the entry and acceptance of qualitative academic research. On the one hand, the first generations of Ph.D. graduates from colleges of education began to study learning from the point of view of teaching and curriculum, thus moving away from studying mental processes that swallowed up research in psychology. This new generation of educator-researchers began to adopt concepts from philosophy and the humanities, such as constructivism, postmodernism, and social criticism theory, to explain the social and political phenomena of education, which quantitative research did not provide. On the other hand, the study of curriculum and teaching caused interest in educational psychology, statistics, and the construction of tests to measure learning to decline. These disciplines disappeared in some Colleges of Education (Walters, 2009).

Between the 1960s and 1980s, educational research began to distance itself from the statistical measurement of educational phenomena and from the search for universal laws that would explain learning across classrooms, schools, and educational systems (Codliffe, 2000). There is an emerging interest in investigating education and learning across diverse educational contexts, such as urban and rural schools and students of diverse socioeconomic levels, across elementary, secondary, and college levels, and across subject areas such as mathematics, languages, or physical education. Proponents of qualitative research capitalized on the philosophical and methodological gaps in the quantitative model of research applied to educational problems. Between the 1950s and 1980s, phenomenology, hermeneutics, and symbolic interaction emerged as research philosophies to address educational problems. The philosophical and methodological tenets of this conglomerate of philosophies eventually became known as the constructivist model of qualitative research. Understanding social dynamics, human behavior, and learning from ethnography in social anthropology, phenomenology in philosophy, and case studies in education positioned qualitative methods as an alternative model of educational research. The qualitative model was based on the following postulates.

The Philosophical Foundation of Qualitative Research

It postulates that human beings know their physical and social worlds through human experience and the senses. The human experience is essential to generate knowledge about the physical and social worlds. It rejects that human beings are mere reactors to the influences of their physical and social realities, as assumed by positivist philosophy. It argues that human behavior results from the interpretation that the person makes of that physical and social realities and the meaning he/she ascribes to them (phenomenology). It establishes that when studying human behavior, the cultural and political elements of the social environment where the experience takes place cannot be discarded (symbolic interaction). It rejects the positivist positioning of absolute truths to recognize the existence of the multiple social realities that constitute the field of education. The argument is that human beings live in a physical world, but life experiences constitute their social reality. These realities are nothing other than the interpretations that each human being assigns to these life experiences. In the light of these interpretations, it is possible to understand human behavior and how each person constructs knowledge and social reality. The qualitative model positioned human behavior as proactive to its social existence. This positioning made a lot of sense among educators because it was consistent with their experiences in the classroom. These positions bring the following elements to educational research scenarios. (a) Social reality is learned and constructed in the social interactions we have with our physical and social worlds. (b) Social reality is complex because it is subjective and relative. Subjectivity and social relativity are tied to the interpretation that each human being gives to his or her experiences. (c) Social knowledge is contextual. Knowledge arises from the situations that restrict the context of the experience.

Qualitative Research Method

The central argument of the qualitative model consisted of the need to "understand" how the actors in the educational system interpret their school, or life, reality to understand how they behave in the face of it. Understanding the phenomena of education through the eyes of those who experience the educational systems made it possible to investigate the range of subjective elements in the field of education and to understand them in the educational context in which they occur. Understanding is important for intervening and correcting learning. This philosophical positioning established methodological positions in juxtaposition to quantitative research methods. These stances are as follows (e.g., Ponce, 2014b; Merriam, 2009; Lichtman, 2006; Woods, 1996; Bogdan and Biklen, 1992; Manen, 1990; Straus and Corbin, 1990):

> **Field research**. As there are multiple social realities in education, the researcher has to enter the educational scenarios to observe these phenomena first-hand, study them, and understand them. Direct observation of situations, behaviors, attitudes, opinions, values, and institutional norms becomes sources of data in the context in which they occur and in qualitative research. To segment, this is to leave out of the study important information for understanding a phenomenon.
>
> **Inductive and holistic research**. The qualitative researcher enters educational scenarios to study a phenomenon starting from an initial concern (inductive) and investigates it progressively until being understood (holistic). Only in this way can he/she know and generate a coherent and contextual explanation of the phenomenon studied and take this knowledge to other application scenarios.
>
> **Emic and etic research**. Since reality is a social construction, the researcher enters the field and draws on the experiences of those directly involved with the phenomenon under investigation to understand it

through their own eyes. In this process, the researcher organizes the evidence he gathers from each participant in the study, from the documents he tasks, the school structures, and the situations he analyzes until a picture of the phenomenon is formed. As a result of the study, the researcher can explain the phenomenon investigated from the point of view of those directly involved (emic perspective) and his or her point of view as a researcher (etic perspective).

The researcher is a research instrument. As education has so many social and subjective dynamics, there is no standardized instrument capable of collecting so much variety and relativity. The researcher is the only instrument capable of managing so much variety and responding to unexpected elements in the study. In qualitative research, the researcher is the one who observes (field observations) and asks questions until he/she understands (through intensive and open interviews).

Flexible and evolutionary research. The researcher may enter a school with a notion of the phenomenon to be investigated and find that it was wrong. He/She then finds it necessary to change his research questions, add other participants to the study, or even change the research technique. It is for this reason that qualitative methodology is flexible and evolutionary.

Research with criterion samples. The qualitative researcher studies particular phenomena related to education and human behavior (criteria). These phenomena do not manifest themselves in the entire population; so there is no need to resort to samples of the universe. For example, if the study criterion is a school dropout, the researcher will resort to dropouts only because that is the "population" where the phenomenon under study manifests itself.

Research that generates qualitative data. Qualitative studies use field observations, open-ended interviews, and document analysis as research and data collection strategies. The data consist of verbal descriptions of what is observed, what people say, or the messages in the study population's documents. These data are not always quantifiable, nor are they always desired to be quantified because the objective is to obtain information that describes the phenomenon under investigation in detail—the greater the report's description, the greater the richness of the data collected.

Strengths and Criticisms of Qualitative Educational Research

In the 21st century, qualitative research is projected as a social critique research model that seeks to contribute to the world's social problems (Denzin and Lincoln, 2011; Flick, 2016). Three research statements are identified in this positioning to contribute to a better society (Flick, 2016):

> **The study of social problems and inequality.** Qualitative research needs to be critical in exposing and explaining the social and political issues being studied. Qualitative research needs to identify vulnerable groups in society, define the problems that affect them, and analyze how institutions deal with these problems. It also needs to formulate recommendations on how these problems can be solved. In this way, qualitative research will remain relevant and valuable to society.
>
> **The scientific study of education.** In education, there is a need to question research practices in the context of the evidence-based research movement, the prevalence of experimental research as the only form of investigation of causal relationships, and the fascination with mixed methods. There is also a need to be critical of the qualitative research methods used as scientific research in education.
>
> **Questioning partisan politics in science.** The objective of questioning practices under the name of scientific research in education, which includes qualitative research, is not to discontinue some forms of research to make way for others. It is to question premises to better understand the methods, the data generated, produce better reports, and politically eliminate the marginalization of qualitative research from the participation of government funds.

In education, the greatest recognized strength of qualitative research is its flexibility to capture the complexity of education and its multiple relationships (Cooley, 2013; Galán, Ruiz-Corbella and Sánchez Melado, 2014; Ponce, 2014a; Ponce and Pagán-Maldonado, 2016c). The major criticism has been the quality of the evidence it generates due to methodological deficiencies. According to Cooley (2013), qualitative education research has provided simple answers to complex questions.

Methodological Evolution

Philosophical research models constitute the first scientific research efforts in education. Quantitative and qualitative research models aim to study, understand, describe, and explain the phenomena of education. Both quantitative and qualitative research explains how educational phenomena affect educational systems' development and effective functioning. Both models use the information they generate on educational phenomena to propose strategies for improvement. In this sense, both research models are scientifically accepted and valid ways of studying education and improving it.

Quantitative and qualitative research models differ in their philosophical views and their positions on the nature of education and the phenomena generated in schools and educational systems. From these differences in the philosophical conceptualization of education, each model coherently outlines its research designs, the procedures it follows, and the data collection techniques or instruments it uses.

The insertion of philosophical models of research into educational research, alternative visions of research education, methodological sides, and philosophical debates on how to research education are created. For some, the relativity that philosophically embraces the qualitative research model did not help the development of education because it did not provide precision in its problems and how to handle them. Concepts such as multiple realities, although logical, made the field of education a world of ideas and relativities. Others understood that educational research needed to be approached from the philosophical point of view of positivism to bring uniformity and precision in handling educational phenomena. The argument was and continues to be that education cannot develop if everything is relative. As we will discuss in Chapter 4, from these philosophical differences about the nature of education, the debates with these research methods and their effectiveness originate. This debate is known in the literature as the quantitative vs. qualitative paradigm debate and is explained in Chapter 4.

4

The Debate of Quantitative and Qualitative Paradigms: Alternate Visions of How to Research Education

Omar A. Ponce

The contrasting visions of educational research discussed in Chapter 3 sparked fierce controversy in the 1970s and 1980s between proponents of academic research that produced causal and generalizable explanations of education (quantitative research) versus those who advocated study that allowed for understanding the social and political phenomena of teaching and learning in diverse educational contexts (qualitative research). The 1980s were characterized as the war of the quantitative versus qualitative paradigms documented in the literature (e.g., Eisner and Peshkin, 1990). The debate in academic and policy forums between proponents of quantitative research and qualitative research was referred to as the paradigm war. The two axes of the controversy were the following: the superiority of one research model over the other (Tashakkori and Teddlie, 1998) and the legitimacy of qualitative research in the field of education (Denzin, 2009). The war of quantitative and qualitative paradigms is essential in the development of educational research because a clearer understanding of the complexity of education as a target of scientific inquiry emerges. The shortcomings of monomethodological research in capturing the complexity of schooling also become evident.

The Attack on Quantitative Research

Between approximately 1960 and 1980, the quantitative research model sparked intense discussions and generated detractors because of its inability to provide all the answers to the problems of public education. In the field of research and evaluation of educational programs in the United States of America, the experimental method for the study of causal relationships did not produce convincing data on the effectiveness of programs designed to combat poverty (War on Poverty) and other social problems of the American nation (Greene, 2007). This dissatisfaction became a topic of debate among philosophers of science, researchers, and academics:

"Whereas the physical sciences dealt with a series of inanimate objects that could be seen as existing outside of us (a world of external, objectively knowable facts), (social) sciences focused on the products of the human mind with all its subjectivity, emotions, and values. From this (Dilthey) concluded that since social reality was the result of conscious human intention, it was impossible to separate interrelationships of what was being investigated. The investigator, who participated in and interpreted that reality [...] the investigator of the social world could only attain an understanding of that world through a process of interpretation-one that inevitably involved a hermeneutical method. (Further) the meaning of human expression was context-bound and could not be divorced from context." (Greene, 2007 p. 34).

Proponents of qualitative research found fertile ground for questioning the effectiveness and relevance of the "scientific method" as applied to education. Critiques and debates centered on the following three themes.

The "Subjectivity" Factor in Education

An argument against the quantitative model's validity, effectiveness, and scope emerges from the dichotomy of the objective versus the subjective. The quantitative model emphasized the investigation of "objective and measurable" phenomena and excluded the range of emotional elements that color education. Factors such as culture, values, attitudes, politics, and morality enter educational systems through their students, teachers, administrators, and the same political scaffolding that makes up the educational system. What to do with the subjective aspects inherent to education that affect student learning and school functioning and cannot always be measured or studied objectively? How to deny that even the way a researcher conceptualizes his research problem introduces subjectivity and prejudices into his study? The "subjectivity" element in explaining learning and understanding how knowledge originates in the field of education highlighted the "limitation" of quantitative methods to study human behavior and its values in educational settings. The questioning of the premises underlying the concept of "objectivity" of the quantitative model brought to light its limited scope to enter into that other non-tangible aspect of values and ethics inherent to the field of education. This debate was critical for the proponents of the quantitative model because it highlighted the fragility of the very conceptualization of social reality on which the quantitative model was based.

The "Context" Factor of Education

Another questioning of the quantitative model's validity, effectiveness, and scope emerges from studies on learning. The conflict began with the question of how learning occurs and whether it is contextual or not. The context in which education occurs became a deadly arrow for the positivist philosophy of the quantitative model. One argument outlined in this issue was how to explain that teaching method and practices that have proven to be "effective and valid" with the quantitative research model did not achieve the same merits when used in other schools and school districts. The analysis of research "contexts" emerges as another significant gap in quantitative research. The context of learning, accompanied by the subjective element of human behavior, highlighted the fragility of the premises underlying the external and universal validity claimed by quantitative research with their meticulous sampling and prediction exercises. These internal and external validity criticisms extended to questioning how reliable the quantitative model was in predicting learning. Studies on the academic performance and intelligence rankings of students, groups, and schools unleashed criticisms of quantitative methodology and the role of research in the field of education. These criticisms reached political and legal arenas. The questioning of whether it is possible to measure a child's intelligence and predict chances of academic success, or how much good it does to society to segregate or classify people, groups, and schools by these quantitative classifications, paved the way for the search for other models of educational research. The argument against the quantitative model was that the social world of educational systems consists of situations with observable facts but with subjective motivations. To study these situations and their attributes, it is necessary to interpret the person who lives them and know the inherent values of the group or culture. In both cases, knowledge of behavior seems to be constructed and tied to the person's interpretation of his or her social world (context) and not to the result of scores on a standardized test or measurement instrument. This possibility moved away from all the premises on which the quantitative method was based.

The Debate on What Is Science vs. What Is Research

The debate on the effectiveness and relevance of the quantitative model to respond to the subjective and contextual aspects of education sparked other discussions on the following three topics: science vs. scientist, science vs. method, and science vs. subjectivity.

> **Science and scientist.** The quantitative model equated science with methodology and methodology with techniques and procedures, as occurred in the natural sciences. In the field of education in the United States of America, the mechanical, sometimes simplistic, and almost artificial application of a scientific method to a wide variety of problems made evident the exclusion of the scientific element in this research model (Paul, 2005). What science can there be in a method if it excludes the research dynamo: the researcher with his intellectual curiosity and his desire to get to the root of things? How to understand, know, solve, and develop the field of education when the essence of science rests on the act of consciousness, deliberation, and intellectual rigor applied by the researcher when studying a problem? There is no scientific method if the researcher is not involved in it from beginning to end.
>
> **Science and method**. As the quantitative research model equated science with the technique, questioning its premises and its relevance to educational problems was also debated. When one examines the beliefs of the scientific model of the natural sciences, one identifies logical postulates for the physical world but not necessarily applicable to the social world. The physical world exists externally and independently of the human being. This material world has attributes and properties. Natural scientists can study physical reactions to other physical objects or their methods objectively and transfer their results without fear of being mistaken because they are talking about biological phenomena. For example, one can study the pollution of a river (effect) if it is used as a recreational area (cause) or explore the impact of a user education campaign to reduce river pollution (effect) because the result is measured in the properties of the physical or natural world. When this model is taken to the field of education, we find

opposite and inapplicable premises. For example, the social world is not external to the human mind. Instead, it is a creation of the human mind. So the social world is full of emotional situations and human interpretations of what happens to them. So, what are the attributes or properties of the social world, which can be studied, measured, and uniform to all human beings? In the field of education, it is possible to learn a student's reaction to his physical world (e.g., a ventilated or warm classroom), to study a student's response to other human beings (e.g., whether he likes his teacher), or to review his interpretations of the events that occur to him (e.g., whether math class is fun or not). These three possibilities cannot be treated with a single methodology because the social world has objective and subjective situations.

Science and subjectivity. The quantitative method emphasizes the value of objective research over the subjective and excludes the researcher from the research process so that it is amoral and free of bias. The process or method is fundamental to verify the new knowledge. As mentioned in the previous sections, educational research can address a situation (observable and objective). This can be interpreted in many ways by school members (subjectivity of the truth). Thus, the scientific method alone does not guarantee the accuracy of knowledge if it excludes the researcher from the process. Recognizing the subjective element in human behavioral research implies that the plan and the researcher must merge in the research process to generate new knowledge and validate it. For example, the researcher may identify an interpretation of an event in one member of the school community that is not shared by the rest of the members. How then does he or she establish the validity of this finding? Criticisms of the quantitative model culminate in two events. (a) The modification of positivist philosophy to a post-positivist one. Post-positivist means that its imperfection as a model of educational research was accepted. Still, its essential elements were retained: the hypothetical-deductive method, verification, and the accumulation of verifiable knowledge, guided by the principles of objectivity and neutrality as norms of research (Phillips and Burbules, 2000). (b) The emergence of qualitative research as an alternative model and the adoption of other philosophies for educational research and evaluation (Greene, 2007; Paul, 2005; Guba, 1990).

The Attack on Qualitative Research

Between the 1960s and 1970s, the qualitative research model emerged as an alternative model of educational research. It is evident in the literature that the popularity of the qualitative model stems from the shortcomings of the quantitative model. As qualitative studies gained greater exposure, so did the criticism of those who favored the quantitative research model. In the 1980s, the first criticisms of the qualitative model were the following.

> **Lack of precision**. The argument was that qualitative research investigates abstract concepts and phenomena, describes them, and communicates them in their studies without many criteria to specify what they are talking about. The absence of measurement in qualitative studies makes them imprecise.

> **Fragility in the cause and effect relationships it establishes.** As a consequence of studying human behaviors in contexts and how people interpret and react to their lived realities, many qualitative researchers conclude their studies with discussions of causal relationships. The criticism here is that these discussions present cause-and-effect relationships without measuring the intensity, duration, or directionality of the relationship they establish.

> **Lack of generalization.** The criticism is that it is impossible to generalize the findings of a qualitative study because of the small sample size used. The theoretical generalization argument presented by qualitative studies lacks empirical validity.

The qualitative researchers paid little attention to these criticisms, stating that they came from people who evaluate the qualitative model with the criteria of the quantitative model. For the qualitative researcher point of view, this assessment is incorrect because the qualitative model operates from different premises. Therefore, they dismissed these criticisms as a reflection of their proponents' poor understanding of the qualitative model because each of them is addressed as follows (Hammersely, 2008).

> **Accuracy.** An excellent qualitative study provides a detailed, rich, and profound description of the phenomena it investigates. These descriptions offer observable indicators of their manifestation (thick report and analytic density). Description and not the measurement is the essence of qualitative research.

Causality. An excellent qualitative study presents detailed descriptions of how a person reacts and acts in light of how he or she interprets his or her life reality. This cause and effect relationship is expressed in terms of the person's process and the consequences of his or her actions. This relationship is observable and not necessarily measurable in concepts such as intensity, duration, and directionality. This type of causality is not the objective of the model.

Generalization. The phenomenon of generalization for qualitative studies does not rely on the size of the sample but the nature of the phenomenon being studied (theoretical generalization). For example, the phenomenon of dropout is similar between school districts and countries. What may change are some cultural elements. A typical case in education on generalization comes from the learning studies that Jean Piaget conducted with his three children. Piaget's learning theory is a solid, popular, and widely used theory in education and was based on his observations of his three children. Qualitative researchers argue that there are elements that cannot be empirically verified, and this is a reality inherent in the study of human beings and their subjectivities.

In the 1980s, despite these criticisms, qualitative research achieved significant popularity in education, both in the United States of America and in European countries (Hammersley, 2008). For some, the qualitative model was positioned as a superior research model to the quantitative model (e.g., Bergman, 2008), while others saw it as an alternative model (e.g., Hammersley, 2008). In the United States of America, it is argued that qualitative research became the dominant model of educational research to the point of displacing the teaching of quantitative research from the curriculum in some Colleges of Education universities (National Research Council, 2002).

Methodological Evolution

Intense academic debates characterized the war of the quantitative and qualitative paradigms of the 1980s. This brought to the surface the lack of consensus and the deep divisions among educational researchers about the nature of knowledge, social reality, and the methods for investigating academic problems (Paul, 2005). It also led to the emergence of a better understanding of the strengths and weaknesses of quantitative and qualitative methods for researching education. With this understanding, the fragility of the validity and limited generalizability potential of quantitative and qualitative research in education became evident.

The debate of quantitative versus qualitative paradigms in 1980, although academic, had political overtones that reflect the contrast between the values of the "scientific community" and the importance of "influential groups" that exert pressure to defend the interests they represent (Lichtman, 2011; Denzin, 2009; Greene, 2007; Paul, 2005). The political presence in education in the United States of America is historical and is evidenced through legislation and the allocation of federal funds for intervention and research programs (Condliffe, 2000). The paradigm wars aroused great concern in political and legislative forums with the quality and usefulness of the research generated in education. It also had the effect of creating influential groups for and against quantitative and qualitative research. The political consequence of the paradigm war emerges with the science-based research movement to bring about evidence-based professional practice, which we will discuss in Chapter 6. This question is still valid at the beginning of the 21st century and has nothing to do with research methods. The concept of the complexity of education also emerges. How a field seems so simple in theory can be so different in practice (Ponce, 2011 and 2014; Ponce, Pagán-Maldonado, & Gómez Galán, 2017 and 2018). The paradigm debate causes alternate philosophies of educational research to emerge. This is the subject of Chapter 5, the emergence of mixed methods in educational research.

5

The Emergence of Mixed Methods in Educational Research

Omar A. Ponce

During the 1970s and 1980s, the paradigm debate between quantitative and qualitative research generated much discussion that contributed to the development and sophistication of educational research. By the 1990s, mixed methods research emerged as a third model of academic research. In this decade, the need for a research methodology that would reconcile the precision of quantitative data with the descriptive richness of qualitative data was evident. Mixed methods research is positioned as a respected research model in education and an accurate model for those who do not wish to conduct quantitative or qualitative research (Caruth, 2013). Studies where quantitative and qualitative research approaches are intentionally integrated as part of the study design are referred to as mixed methods research. Several movements contribute to the development of mixed methods research in the field of education (Tashakkori and Teddlie, 1998; Greene, 2007; Ponce, 2011, 2014 and 2016): the emergence of the concept of triangulation, the rejection of paradigmatic puritanism, a pragmatic view of research, and the thesis of compatibility between quantitative and qualitative methods. The emergence of mixed methods constitutes a significant methodological evolution in educational research.

Evolution of the Triangulation Concept

Triangulation means the possibility of having more than one data collection technique in a study as a strategy to corroborate findings, for example, using a questionnaire and an interview in the same research to collect data on the same phenomenon. The concept of triangulation emerged in the 1950s due to negative criticisms of experimental research in the social sciences and education (Greene, 2007; Tashakkori and Teddlie, 1998). The discussion of paradigms and the search for strategies to raise the quality of educational research produced broader visions of the concept of triangulation, such as the possibility of bringing this idea to research design, for example, using a quantitative and a qualitative study simultaneously to study a phenomenon. Mixed methods research emerges from the concept of triangulation (Greene, 2007). The contradictory findings of qualitative and quantitative studies opened a window of possibilities to investigate these, but resorting to practices, strategies, and designs not typical of these research models. The need for tasks where the data generated are confirmed or validated by more than one mechanism highlighted the concept of triangulation as central to the new studies to be developed (Berman, 2008).

Rejection of Paradigmatic Puritanism

Between the 1960s and 1980s, prominent social and educational researchers moved away from the paradigmatic puritanism that characterized the paradigm wars. Paradigmatic puritanism is the term used to describe the stance of identifying oneself as a quantitative or qualitative researcher. These educational researchers begin to combine and integrate philosophies, methodologies, and data in their studies (Greene, 2007). They defend the "compatibility" thesis between quantitative and qualitative models (Tashakkori and Teddlie, 1998). The compatibility thesis argues that quantitative and qualitative models complement each other and that both are important and necessary in educational research and evaluation (Tashakkori and Teddlie, 1998). The paradigm war made many evident philosophies and forms of academic research, all valid and essential. Most of the discrepancies between quantitative and qualitative researchers are more philosophical than methodological (Paul, 2005).

The Emergence of Pragmatism as a Research Philosophy

With the rejection of paradigmatic puritanism, pragmatism emerged as a conciliatory research philosophy between quantitative and qualitative research. Under pragmatism, mixed educational researchers/evaluators took shelter. Mixed methods research begins to define itself in practice and name. This evolution was observed in the following transitions (Tashakkori and Teddlie, 1998): (a) **multi-technique research**—studies where more than one research technique was used as a data triangulation strategy; (b) **combined methods research**—studies where quantitative and qualitative approaches are employed as components of the research design are connected or integrated; (c) **mixed methods research**—studies where quantitative and qualitative models are intentionally integrated into their philosophies, methodologies, and data.

The Thesis of Compatibility Between Quantitative and Qualitative Methods Research Emerges

The thesis of compatibility between quantitative and qualitative research gained acceptance in the United States of America and European scientific research community. Quantitative and qualitative research, rather than different research approaches, are understood as complementary to each other (Phillips, 2009). (a) The paradigm war made it clear that neither of the two existing models of educational research, quantitative and qualitative, was superior to the other because both have strengths and limitations. Monomethodological research cannot claim knowledge or absolute truths because of the inherent limitations of their respective models. (b) The influence of politics in federal education in the United States determines allocating funds for intervention and research programs. The paradigm war evidenced that quantitative studies respond well to those administrative spheres that need to report accurate intervention data due to the allocation of funds to projects and programs. Qualitative research appeals to educators and those engaged in direct intervention with students and parents. Many federally funded education research projects understood these dynamics. They began to use teams of qualitative and quantitative researcher-evaluators to generate integrated data so that they could compete for these federal funds in favor of the federal administration but with the support of faculty and other staff directly involved in the intervention work. Many researchers saw working in these teams as great opportunities for professional growth (Berman, 2008). The need to integrate both models into one seems to be the logical step in the face of educational problems tinged with objective and subjective elements (Caruth, 2013).

In 1980, the number of published studies employing qualitative and quantitative methods as components of the research design increased. The argument of these studies to integrate or combine quantitative and qualitative models in the design of their studies was the methodological robustness to study the problems of education (Caruth, 2013; Greene, 2007; Tashakkori and Teddlie, 1998): (a) the need to have more complete, rich, and valid studies than a monomethodological study by having more than one way to approach and collect data on the phenomenon under investigation; (b) the need to have more complete, richer, and more reliable data allows a better understanding of the phenomenon under investigation; (c) the need to use one research approach to support the other—the methodological strength of one approach compensates for the deficiency of the other; (d) the need to explore those new perspectives that emerge when quantitative and qualitative data are integrated into one cannot be generated separately.

Mixed Methods Research

Mixed methods research in education is grounded in pragmatist philosophy (Campos, 2009; Teddlie and Tashakkori, 2009; Ponce, 2011, 2014, and 2016). Pragmatism postulates that knowledge, human behavior, and social reality are inseparable entities or domains intimately related in a continuum (Biesta and Burbules, 2003). These are not independent entities as positivist philosophies, phenomenology, and symbolic interaction, where it is assumed that social reality and knowledge are separate entities. Knowledge cannot be studied, understood, or applied without an external reality that serves as a context. Social reality cannot be explored or understood if human beings do not interact and experience it. It is on this point that pragmatism differs from other philosophies of educational research.

Pragmatists argue that knowledge begins to be generated when human beings act on their environment. This interaction comes because they interpret it or because they experience needs or interests they seek to satisfy. Humans act based on what they think they should do (reasoning) to meet these needs or interests. The thinking process can be very subjective, but the consequences of actions are objective when they become individual or social behaviors. From this point of view, knowledge is experiential and evolutionary since it is constructed and refined by the actions and experiences of human beings with their social and physical worlds. Therefore, human expertise is imperative to understand how knowledge is generated. Human beings learn from their actions and the consequences they bring about. The process through which an idea passes, from thinking it to converting it into action and living its consequences when inserted in the social world, constitutes the focus and structure of analysis to understand knowledge (sociology of knowledge). Knowledge is generated in a three-phase system: (a) thought or idea, (b) action or behavior, and (c) consequence. Observable indicators of human behavior that can be linked to thought can be generated from actions and consequences. Knowledge has the following four characteristics. First, it is contextual because it is constructed from experiences with specific situations.

Second, it has subjective (the person's intentions) and objective (the facts or consequences) aspects because it arises from the human experience with its environment's physical and social worlds. Third, in the quest to understand how knowledge is constructed, human experience's political, ethical, and moral aspects cannot be discarded. Fourth, the validity of knowledge is determined in the light of its consequences, that is, whether it brings about the desired social good.

From the perspective of pragmatism, social reality is external to human beings but intimately related to human experience. Reality is composed of three objective and subjective elements: (a) the physical world where human beings inhabit and interact (e.g., whether it is hot or cold, or whether it is day or night); (b) the social interactions between human beings (e.g., the teacher–student relationship); (c) the appreciation of the reality that human beings make, the consequences of their actions, and whether they bring the result they expect (e.g., if the student perceives the class as boring and decides not to pay attention). Although knowledge is formed in the human mind, it is not a mere product of a world of ideas but of the interaction of the human being with that external social reality. Knowledge is generated by the mind and becomes, at the same time, information about the social world. From this perspective, social reality is never static. To the same extent that human beings interact with their social reality, they develop understanding and mastery. This can lead to changes in needs, interests, and behaviors. For example, as a child masters the basic skills of a sport, he or she may become interested in experimenting with other more complex skills that stimulate or challenge him or her.

Suppose the assumptions about the objective and subjective nature of knowledge and its contextual origin are accepted as valid. In that case, qualitative research's phenomenology and social construction are relevant to understanding the emotional aspects of education. It also means that the actors' experiences, when converted into actions, manifest observable indicators with their consequences that can be specified, measured, and corroborated. This makes the measurement and numerical precision of quantitative studies relevant. Educational research has to be mixed if it is to capture the extent and complexity of education. Pragmatism applied to research positions the researcher in need to study the objective and subjective aspects of the phenomena he/she is investigating to understand, explain, or solve. Hence, the need to combine or integrate quantitative and qualitative approaches to answer the questions emanate from the objective and subjective aspects of human experience. The approach to complex problems is always progressive, in stages or by levels of understanding. It is rarely possible to understand them with a single study and a single method. To the extent that the researcher enters into the complexity of a problem, he/she gains an understanding of it. Each effort to go deeper into the problem helps to refine further the knowledge that is achieved. The more experience a study produces, the greater its effectiveness and the greater the likelihood of solving it. A researcher cannot, and should not, deny the complexity of educational scenarios if he or she wishes to generate studies that contribute to the solution of academic problems. Methodologically speaking, it is difficult to establish an anticipated and prescribed formula of when or how to combine or integrate quantitative and qualitative approaches to investigate a problem. This decision emanates from the study's research questions, the particularities of the situation to be investigated, and the researcher's curious and creative desire to get to the root of things. In terms of methodology, this means that researchers will decide how and when to combine/integrate quantitative and qualitative research approaches in light of the research objective they are pursuing. The more a combination of quantitative and qualitative research approaches brings the researcher closer to his or her purpose or to the particularities of the phenomenon/problem he or she is investigating, the greater his or her effectiveness as a researcher and the greater validity the study will have. This is called a pragmatic decision. The above means that the researcher should pay close attention to the planning phase of the study to achieve an effective integration or combination of quantitative and qualitative approaches, which allows him/her to handle the inductive flexibility of the qualitative and the deductive structuring of the quantitative.

Strengths and Criticisms of Mixed Methods Research in Education

The strength of mixed methods research in education is the possibility of combining or integrating quantitative and qualitative methods in the study of a problem and the scope. This has to capture the complexity of schooling (Creswell and Garrett, 2008; Ponce, 2011; Ponce, 2014c; Mertens, 2015; Ponce, 2016; Ponce and Pagán-Maldonado, 2016d). This possibility includes studying causal relationships (Ponce, 2011 and 2014d; Ponce and Pagán-Maldonado, 2015; Ponce, 2016; Ponce and Pagán-Maldonado, 2016d). According to Ponce and Pagán-Maldonado (2015), the complexity of education emanates from the phenomena under investigation (e.g., teaching and learning processes), from the educational contexts (e.g., the culture of a school and its policies), and its administrative structures with multiple levels of functioning (Ponce, 2016). This same attribute is recognized in mixed methods research in other academic disciplines (i.e., Mertens, 2015; Jokonya, 2016). Mixed methods provide the possibility of measuring phenomena occurring in education and in the schools' culture and policies where they manifest themselves. In educational research, triangulation and parallel-phase complement designs have been successfully used to measure causal relationships related to teaching and learning processes and environmental factors and conditions with occupational stress (Ponce and Pagán-Maldonado, 2015). Multisample and multilevel designs provide to approach the diversity of educational dynamics of interrelationships of individuals or groups or that are products of the multiple structural levels of educational systems (Ponce, 2011; Ponce, 2014c; Ponce and Pagán-Maldonado, 2015; Ponce, 2016; Ponce and Pagán-Maldonado, 2016d).

The major criticism of mixed methods research is its complexity. It involves combining or integrating quantitative and qualitative research methods to investigate a problem to capture its full extent, scope, and complexity. This requires mastery of both quantitative and qualitative research methods. Mixed methods research is time-consuming in education (Ponce & Pagán-Maldonado, 2015) and other academic disciplines (i.e., Jokonya, 2016). The complexity of mixed methods has raised the question of whether this research model turns out to be more appropriate for teams of researchers than for individual researchers (Creswell, 2016).

Methodological Evolution

Mixed methods were widely embraced by educational researchers and political and influential research funding groups. In academia, a growing number of quantitative and qualitative scholars and researchers took up the defense of mixed studies because they conduct this type of research or because they recognize the value of this model. It is known in the literature as the complementarity thesis. The complementarity thesis argues that the difference between qualitative and quantitative approaches exists more in the minds of researchers, at the philosophical level than, in practice, at the methodological level. In practice, both models complement each other because the strengths of one research approach compensate for the shortcomings of the other (Ponce and Pagán-Maldonado, 2015). One position is to argue that what is relevant in educational research are the research questions, the generated data, and how these enable the profession to advance

(Greene, 2007). The characteristics of the problem determine the method, and this has nothing to do with philosophies (Campos, 2009). According to Phillips (2009), mixed studies constitute a research model that reconciles the elements of scientific research and the answers to the methodological deficiencies of quantitative and qualitative models of educational research. The model also satisfies the political sector that vetoes qualitative education research. This group recognizes the challenge of integrating qualitative and quantitative approaches as components of a study if they stick to the traditional philosophy-methodology structure. The emergence of mixed methods in educational research did not completely mitigate the paradigmatic and philosophical debates among educational researchers about how to research education. Although mixed methods research is a conciliatory point between the objective elements of quantitative research and the subjective elements of qualitative research that were debated in the paradigm wars, the fact remains that mixed methods research is a philosophical model of inquiry and one that does not enjoy favor with some qualitative educational researchers (Flick, 2016).

6

Science-Based Research and Evidence-Based Decision-Making

Omar A. Ponce

Between 1994 and 2001, discussion about the quality and effectiveness of public education were revived in political forums. This time, the topic of quality and the usefulness of educational research is linked to the quality of public education in the United States of America (Ponce, 2014 and 2016; Labarre, 2004; Condliffe, 2000), in England (Pring, 2000; Hammersley, 2007) and in Australia (Erickson, 2011). From the policy point of view, it became clear that the expectation with educational research is that it provides the solutions to the problems of public education and prescribes its practices to develop scientifically. From the point of view of educational researchers, it became evident that the search for that ideal scientific method, which has been the dynamo of its development, has also been the axis of controversies with the quality and usefulness of educational research (Condliffe and Shulman, 1999; Condliffe, 2000; Labaree, 2004; Walters, Lareau and Ranis, 2009). The debate of quantitative and qualitative paradigms in the 1980s was essentially more philosophical than a methodological one. In 2002, the Federal Government of the United States of America politically ended the controversy by regulating educational research seeking federal funds (Shavelson and Towne, 2002). A similar situation occurred in England (Hammersley, 2007). The political action consisted of the delegitimization of qualitative research by not considering its scientific research and the imposition of a science-based research model to bring about science-based evidence-based education, as occurs in the field of medicine. Denzin (2009) uses the term "the second paradigm war" to describe the political debate sparked by the Federal Government's intervention in regulating educational research. Barnhouse, Lareau, and Ranis (2009) use the term "the political judgment of educational research" to refer to this action by the Federal Government. The controversy is whether educational research can emulate research that occurs in the medical field. The discussion of this second debate charts the course for the development of academic research as we enter the 21st century. This chapter discusses the political action of imposing a science-based research model to bring about evidence-based professional practice as occurs in medicine. This discussion brings to light again the issue of the ineffectiveness of educational research in improving education.

Criticism of the Effectiveness of Educational Research

The discussion on the ineffectiveness of educational research to improve public education revolved around three issues: the usefulness, quality, and scope of scholarly research.

Criticism of the Usefulness of Educational Research

Educational research has developed in two categories: basic research and applied research. Basic research refers to academic studies developed in universities and published in dissertations, textbooks, and articles in prestigious scientific journals. Basic research has focused on the study of education as an academic discipline. The objective has been to produce new knowledge that contributes to understanding education and its development through theories and models that guide practice. Applied research refers to studies developed in public schools to understand their functioning and specific problems. This type of research emerges with the funds that the federal government makes available to investigate public education issues. The tendency has been that the federal government identifies the problems it is interested in exploring. The usefulness of basic research and applied research has been criticized by the federal government of the United States of America. Basic research is criticized because it is understood as theoretical research, unconnected to the real problems of public education, and too technical for the understanding and practical use of the classroom teacher. Applied research has been criticized because it does not provide the solutions to the problems of education. The effectiveness of both forms of research has been criticized because they do not contribute to educational policy development and decision-making on how to improve education (Shavelson and Towne, 2002).

The critique of the utility of educational research implies educational research that is limited to problem-solving. This excludes studies conducted to explore and illuminate the complex problems and dynamics that have brought depth and scope to education. From the perspective of research utility, it is clear that some methods are more relevant than others (Ranis, 2009). Most of the studies that can be classified as primary research are those carried out by university professors and that respond to the evaluation criteria imposed by universities for their hiring or renewal of contracts (Hammersley, 2007). However, not all educational research is oriented to produce educational policies (Ranis, 2009). Indeed, the field of educational research can be more responsive to the needs of public education and be more concerned with the quality and usefulness of the study it generates (Vinovskis, 2009). The questioning of the use of educational research should be seen as an opportunity to directly link the production of new knowledge with the usefulness of knowledge (Ranis, 2009).

Criticism of the Quality of Educational Research

Since its inception, the expectation with educational research has been that it prescribes what education should be like to make educational systems effective in training new generations of citizens and professionals. Educational research has succeeded in generating valid explanations and impressive advances in education. However, it has not been able to control or predict education phenomena to anticipate events, control situations, or predict outcomes. For example, in education, it has not been possible to identify the universal method that guarantees that all students entering an educational system graduate with the academic development established by the educational goals and objectives. Nor has it been possible to identify, anticipate, or control students dropping out of school. The criticism is that much of the scholarly research is not linked to the classroom practice of public systems or their educational policies. This criticism emerges when comparing educational research to research in medicine or agriculture. This argument has been used to question the quality and scientificity of educational research in capturing the complexity of education (Walters, 2009; Hammersley, 2007; Pring, 2000; Carr and Kermmis, 1985).

Studies have shown that the quality of basic educational research compares favorably with research in other disciplines that study human behavior (Walters, Lareau, and Ranis, 2009). According to Walters, Lareau, and Ranis (2009), the controversy with the quality of educational research should be with applied research because it does not provide those "solutions" that the federal government expects from research. For some policy makers, administrators, and educators, educational research must generate the data to prescribe classroom educational practice and develop their educational policy (Shavelson and Towne, 2002). Defining educational training or its policies is understood as an unrealistic expectation if one knows the nature of education and the data generated from this phenomenon. This does not mean that educational research can inform the practice of the profession and its educational policies (Hammersley, 2007; Ponce, 2016; Ponce, Pagán-Maldonado, & Gómez Galán, 2017 & 2018).

Education consists of the ideas, knowledge, values, and skills taught (transmitted) to students through lectures, discussions, field trips, readings, and assignments, to mention only a few. The student goes through these educational experiences or curricula. It is sought that they process them in their minds, experience them through real educational experiences, or relate them to their life realities. Although teaching is an action intended to achieve academic goals and objectives, its effects occur in students' minds and are manifested through feelings, opinions, aptitudes, and attitudes. Teaching and learning are known as symbolic interactions between educators and students (Hammersley, 2007). Much of the data generated from symbolic interactions can be observed with the eye but are challenging to pinpoint, measure, and corroborate because they emanate from intangible aspects of the human being. For example, verifying a student's physical changes is simple because it is measured by whether he/she grew X number of inches in height or gained Y number of pounds in body weight. This type of data is easy to show because it corresponds to the student's physical body. The evidence of cognitive, affective, or psychomotor changes as a consequence of the education received by the student becomes a challenge for educational researchers because they correspond to mental processes, and their effects are manifested in the soul, in the character of the person, and the social world of the student. These changes can be observed and appreciated, but they are difficult to pinpoint, as it would be to measure the temperature of a classroom. For example, how do you measure whether one person is more moral than another or happier? Likewise, it is a tremendous challenge for educational researchers to explain whether changes in students are the direct result of the curriculum, teaching methods, the management style imposed by the school principal in the educational institution, or the influence of student peers. Added to this is that students are constantly developing; so, today, they may be at one level of maturity and two weeks later, reach another level of development (Shavelson and Towne, 2002). The ambivalent and difficult to measure and corroborate nature of the data generated in education makes educational research a slippery science for reducing this reality to a number (Condliffe, 2000; Labaree, 2004). For

Labaree (2012), the overemphasis on quantification that has occurred in education has been to its detriment because quantitative data reduces facts and hides realities by reducing the profession to numbers.

The data generated from education are not only ambivalent but also fragmented. The symbolic interactions that characterize the field of education have the distinction of being multidimensional. This means that the same phenomenon can be understood from different optics and equally valid (Hammersley, 2007; Shavelson and Towne, 2002). For example, the effectiveness of teaching can be evaluated from the perspective of the teacher who performs it, the school principal who supervises it, or the student who receives it. Teaching can be evaluated from the point of view of the foundation from which it is developed: its planning, its sequence of steps, and how much it resembles what is established by institutional policy. Teaching could be evaluated from a pragmatic perspective of whether it produces the ideals found in the goals, objectives, and educational standards. The challenge with the multidimensional phenomena of education is to generate that unique and great explanation of how it is to guide educational practice. This fact places the educational researcher in the dilemma of identifying from which perspective it is best to approach the phenomena he/she studies and thus recognize the limitations that his/her research entails since he/she cannot claim absolute truths. This has generated two controversies about the quality of educational research. The first controversy is between those who argue that the researcher is obliged to identify the unique optic or absolute truth that explains the phenomenon (post-positivist philosophy) and those who understand that the researcher can only capture aspects of the phenomenon, according to the optic from which he/she develops his/her study (phenomenology and symbolic interaction). In other words, there are politicians, researchers, and educators who argue that educational research can produce absolute truths as in medicine or agriculture and others who think it is impossible. This causes educational research to be understood as an imprecise science (Shavelson and Towne, 2002). The second controversy centers on the relevance of quantitative, qualitative, and mixed methods to study education.

The solutions to many problems in education are practical. For example, what should the teacher modify in the classroom if the student is not learning at the pace he or she should or what type of discipline should be used if the students do not want to read. The symbolic and multidimensional nature of education challenges the scope of data generated from a study because one is working with people who interpret the realities that live in educational settings (Lingenfelter, 2011; Hammersley, 2007; Shavelson and Towne, 2002). Many problems in education can be solved in different ways because there is no single solution. For example, some parents prohibit their children from watching television until they complete assignments to concentrate, while others allow them to do assignments in front of the television. Each has different tactics and valid arguments for getting their children to complete assignments at home. Each achieves the same goal of getting their children to complete assignments with different alternatives to the same problem. This has generated controversy about the role of research in the field of education: (a) some understand that the role of educational research is to generate the knowledge to understand education so that educators and administrators are informed and reach consensus on how they can best do their work in light of that scientific understanding (Biesta and Burbules, 2003); (b) the federal government advocates for scientific research that prescribes solutions for teachers to achieve evidence-based professional practice, as has occurred in medicine and agriculture (Shavelson and Towne, 2002). Faced with this dilemma, educational researchers face generating data that help to understand education or solve its problems by prescribing solutions.

Criticism of the Scope of Educational Research

Contemporary education is organized by public and private educational systems ranging from nursery schools to universities. Each of these institutions pursues the development of students. As a standardized activity, education is developed through diverse study subjects in primary and secondary schools and a range of professional disciplines in universities. The management and structures for organizing and administering these subjects, curricula, and programs of study represent another development in education. Contemporary education presents a complex web of human interrelationships and educational activities—practices—experiences that interact toward achieving academic goals and objectives. These interrelationships are framed in a range of educational institutions with diverse administrative structures and philosophies. Educational research has been entering into the complexity of education through research by disciplines of study and institutional research.

Research by the discipline of study means using systematic methods and procedures applied to the study of specialties—for example, research in physical education, special education, or early childhood education. The purpose of discipline-based research is to improve the professions that make up the field of education. Institutional research means using these scientific research methods and procedures applied to the study of the functioning and effectiveness of educational institutions. The purpose of institutional research is to evaluate educational institutions to generate data that contribute to administrative decision-making and the systematic planning of their development projections. This has resulted in criticism of educational research for not capturing the problems of education in their totality. The argument is that educational research has been effective in portraying its educational subjects and educational institutions but not the effectiveness of the country's education (Walters, Lareau, & Ranis, 2009; Labaree, 2004; Condliffe, 2000; Condliffe & Shulman, 1999; Carr & Kermit, 1985). This criticism is the result of two factors. The first factor is the absence of a national research plan that defines the problems of education. This has left it to the discretion of researchers to address those problems documented in the literature or identified by educational administrators (Hammersley, 2007; Shavelson and Towne, 2002). This fact is highlighted each time a change in government and educational needs and problems are redefined (Hammersley, 2007). The second factor is that the absence of a shared vision of educational research lies in the difficulty of defining what education is to research it uniformly (Labaree, 2004; Pring, 2000; Carr and Kermmis, 1985).

The Debate on the Effectiveness of Educational Research: The Second Paradigm War

Two controversies emerge with the political imposition of a science-based research model to provoke an evidence-based professional practice: the logic that justifies such an imposition and the relevance of the science-based model to provoke an evidence-based professional practice in the field of education. The second paradigm war was unleashed with the publication of the National Research Council (2002) (Denzin, 2009) and continues at the publication of this book. The NRC is a non-profit entity (Shavelson and Towne, 2002), subsidized by the Federal Government of Education of the United States of America (Denzin, 2009). Its role is to recommend policies to resolve research controversies (Shavelson and Towne, 2002). In its publication, *Scientific Research in Education* (2002), the NRC takes the following position on educational research:

1. The quality of educational research is questionable. There is no consensus on the criteria for defining the quality of educational research.

2. Educational research is fragmented and without much connection to inform aspects of public policy or classroom practice.
3. There are too many philosophical and methodological differences among researchers about what educational research should be like.
4. The tendency in schools of education to favor qualitative research, in some cases, at the expense of quantitative research, opened the door to criticism about the quality of educational research.
5. The distinction made by academia between quantitative and qualitative research, and the categorization of educational research as essential and applied, is erroneous. Scientific research is the same in any discipline of study.

To improve the quality of educational research, it recommends that all federally funded academic research be scientific so that teaching practice is evidence-based. It asserts that the policy community wants to invest money to improve education safely. It establishes the following criteria for considering research to be scientific:

1. The use of systematic and rigorous procedures that allow direct investigation of the research question.
2. Recognize the context of the research.
3. That it responds to conceptual frameworks that allow to see the relationship of the study with the research logic, its data, and interpretation.
4. To test and corroborate hypotheses and theories.
5. Allowing the data to be related to the method so that the findings can be verified and are cumulative.
6. It rejects postmodernist schools of thought in social research which argue that knowledge is not objective. It indicates that the social world can be described scientifically and reliably. It uses, as an example, the fact that several researchers reach a consensus on what they observe. Although he recognizes scientific, educational research designs, such as control group experiments in program evaluation, and observational and in-depth interview studies such as ethnographies and case studies, not all academic research satisfies criteria 1−5 listed above.
7. Recommend implementing institutional guidelines and compliance committees to evaluate, ensure, and raise the quality of federally funded research.

Phillips (2009), Vinovskis (2009), and Walters (2009) trace the influences of this policy intervention to the first paradigm war of the 1980s. Walters

(2009) notes that the debate about educational research, since 1995, does not focus on the superiority of quantitative or qualitative methods. The discussion addresses the issue of the quality, utility, and role of educational research in society. This must be understood to outline strategies and counter the attack on educational research:

1. It is questioned whether educational research is a reliable science or whether it is a science.
2. Their usefulness for educational development, classroom practice, or educational public policy development is questioned.
3. Educational research is criticized for its lack of generalization and defined theoretical frameworks to inform teacher practice.
4. Its methodological ambivalence, the research questions it addresses, and the discrepancies between the research philosophies employed and the designs that guide them are criticized.

The first war of the quantitative versus qualitative paradigms facilitated three parallel movements that conclude in the "political judgment" facing educational research (Walters, 2009).

> **The evidence-based professional practice movement.** The evidence-based practice movement arose from the neoliberal policies that occurred worldwide in the 1980s. The argument was to have transparent governments for constituents in governance and financial management. Although these policies began in government, they took educational reforms to demonstrate the relationship between government management and public investment (accountability or accountability). The "accountability movement" was first applied to public schools under the concept of quality standards and standardized tests to measure learning (Kincheloe, 2003). It was then taken to universities under different names and approaches (Hammersley, 2007), such as learning assessment and accreditation standards (Austin and Linsing, 2011). In the 1980s, the science-based medicine movement emerged based on the notion of evidence-guided practice. The effect of the evidence-based practice movement was an awakening of the positivist philosophy on educational research and the questioning of the role of research in the social sciences and education. Two positions on the role of research emerge and are debated from this evolution (Hammersley, 2007): (a) **the instrumental role of research—** it emanates from positivist philosophy, where the function of research is

to improve the world by developing new and more effective techniques to replace traditional ones and the constant demonstration of how things work and can be improved; (b) **the informative role of research**— research as an exercise to reflect on social and educational practices beyond their instrumental nature. Research is presented as the search for knowledge, as an exercise of intellectual curiosity, and as a critical exercise of evaluation of current practices, even if there is satisfaction with the status quo of things.

The 1980s also saw the emergence of the questioning of educational quality with the report "A Nation at Risk." The political and academic involvement following this report is documented in several bills (Walters, 2009). Strategies to improve education and intervene with social problems affecting public schools took federal funding for educational programs and social impact projects. Evaluating the effectiveness of these programs/projects and their fiscal investment allowed program evaluation, or research-evaluation, to become a discipline of study. Some departments of education, economics, social sciences, statistics, and research began to offer professional certificates and academic degrees in applied research in educational program evaluation. In the beginning, experimental methods were used, then qualitative methods, and, eventually, mixed methods. This facilitated the emergence of the consulting firm industry, which competes for federal funds to evaluate educational programs. This allowed evaluation research to cross academic boundaries between disciplines. The first paradigm war raised three questions about educational research: its quality and usefulness in solving educational problems, and the role of educational research in guiding classroom practice and the development of educational public policy. The argument generated was that the "supposedly poor quality of public education" is due to the "poor quality of educational research" and its poor relationship and usefulness in guiding classroom practice. Although the quality of educational research is questioned, studies published in professional education journals compare favorably in rigor and scientificity with studies from other academic disciplines (Schneider, 2009).

The resurgence of the science-based research movement. The first paradigm war caused many educators' interest in quantitative research methods to wane. In some Colleges of Education, qualitative research ultimately displaced quantitative research and evaluation methods from the curriculum (Shavelson & Towne, 2002). The entry of social critique and postmodernist research methods and philosophies in the 1990s fueled the questioning of the quality of educational research and its scientificity. Social-critical and postmodernist researchers criticized quantitative methods and constructivist qualitative research. Researchers who embraced the scientific culture of positivist methods in which they were trained as educators, psychologists, economists, statisticians, measurement-evaluation specialists, and business people quickly criticized social criticism and postmodernist studies. Historically, scientific culture in universities has been centered on positivist philosophy. At the same time, in the 1990s, the science-based research movement and evidence-based practice re-emerged strongly. These two groups found political forums with the issue of the quality of public education in the country. They raised the questioning of the quality and scientificity of educational research being debated at this time (Walters, 2009).

The political control of scientific research. The first paradigm war revealed the strengths and weaknesses of quantitative and qualitative methods used in educational research. These methods are used in other disciplines that study human behavior. Although scientific research and politics are two separate enterprises, the history of scientific research has shown that whoever defines what science is controls the scientific enterprise. The political imposition of science-centered research in education reflects the power and control struggle that political groups, scientists, and academics unleashed at the country's highest levels to control the nation's scientific research politically. One view of this dynamic argues that this power struggle aims to manage scientific research and the money that goes with it. This control struggle begins with educational research and will eventually touch other social science disciplines (Walters, 2009). The root of this perspective seems to be linked to the intellectual movement unleashed by a group of prominent academics and researchers to turn social sciences and education into natural science disciplines: evolutionary biology, genetics, and brain sciences. This group argues that research in the social sciences and education is not scientific research. Research in these disciplines is ideological research because it does not have a paradigm to support it. According to this group, what research in the social sciences and education has done is to distort reality and create the illusion that human behavior cannot be studied with the methods of the natural sciences (Klemke, Hollinger, Rudge, 1998). According to Klemke, Hollinger, Rudge (1998, p.106), the argument that this group generated is overdue because of the sophistication that research in these academic disciplines has reached.

The second paradigm war was between educational researchers and educators against groups outside the field of education. The outside groups are composed of university academics from other disciplines of study, business people, and politicians who criticize education and educational research quality. The criticism of these groups comes from the desire to improve public education in the country, the apparent economic interest they derive from federal funds, and the control they exert over scientific research.

It is argued that the second paradigm war is the consequence of the 1980 paradigm war (Lichtman, 2011; Denzin, 2009; Greene, 2007; Paul, 2005). Others argue that the first paradigm war never ended. Simply, the debate shifted in the 1990s to the high political spheres in Washington for airing (Phillips, 2009; Vinovskis, 2009; Walters, 2009). The quantitative and qualitative paradigm wars of the 1980s brought to the surface the lack of consensus and the divisions that exist among educational researchers about the nature of knowledge, social reality, and the methods of how to investigate problems in education (Paul, 2005). The 1980 paradigm debate, although academic, had political overtones that reflect the contrast between the values of the "scientific community" and the values of "influential groups" that lobby to defend the interests they represent (Lichtman, 2011; Denzin, 2009; Greene, 2007; Paul, 2005). The political presence in education in the United States of America is historical and is evidenced through legislation and federal funds for intervention and research programs (Condliffe, 2000). Greene (2007) summarizes the policy debate in the paradigm war of the 1980s as a question of values and utility: To whom does the research conducted in the field of education respond and benefit? Qualitative studies respond to the needs of teachers and other employees who work directly with students and their families because they understand the values and cultures of these oppressed and marginalized groups in a postmodernist society. Quantitative studies respond to the interests of administrators and investors who want to know the results of the funds they invest in special programs and projects to improve education. Educational research becomes political to the extent that its findings help to develop legislation or evaluate its scope. The criticism and marginalization of qualitative research, especially social criticism and postmodernist research, in the allocation of federal funds come from influential groups whose legislation favors some groups at the expense of others (Denzin, 2009). The political message to the academic community seems to be the interest in a uniform view of what "scientific research" means in education (National Research Council, 2002). Paul (2005) identifies the work of the National Research Council and the No Child Left Behind Act (NCLB, 2001) as the two markers of the vision that influential sectors are imposing on educational research in the early 21st century: "science-based research" and "evidence-centered professional practice." In the face of such a plurality of values and research approaches, ethics in educational research emerges as the methodological centerpiece (Paul, 2005). What values will educational research represent: the values of the agencies that fund educational research, the researchers' values conducting these studies, or the values of the people being studied (Mertens, 2005)?

For some, the second paradigm war is political (Denzin, 2009; Walters, 2009; Ellis, 2005). The feeling is that there are people in academia, industry, and the federal government who want qualitative research to disappear (Lichtman, 2011; Denzin, 2009; Walters, 2009). Qualitative research is disadvantaged by not being eligible for federal funding. If one examines the history of qualitative research in education, the infusion of federal funds helped develop this research model. Ellis (2005) points out that this is not the first time the federal government intervenes in educational policy issues. He adds that there are other foundations in the United States of America that fund research that contributes to education development. Denzin (2009) warns of the challenge if other funders adhere to federal regulations. He categorizes political meddling as an attack on democracy and academic freedom of expression, which touches all disciplines of the social sciences, by imposing a model of how educational research should be. He indicates that it is a discriminatory policy because it marginalizes those minority groups that were heard in political forums through social criticism and postmodernist studies. He adds that the NRC created a hostile environment for those who research the qualitative research model. He argues that science-based research is a political attempt to control and manipulate the truth through legislation such as the No Child Left Behind Act and publications such as the NRC. This type of political strategy became evident in public administration as part of the neoliberal philosophies of the 1980s, which many governments around the world adopted. The particularities of these movements were to make decisions in light of evidence to have transparent governments in the management of public finances. The political aspect of the science-based research model emerges from imposing a model of how educational research has to be and what is considered valid evidence for the Federal Government. These types of measures, rather than regulating, are adopted to exclude. The application of this legislation in educational research means excluding qualitative research, especially postmodernist and social criticism research.

The objective of evidence-based management is transparency. The result of unilateral policies translates into favoring one group of political and social influence at the expense of others who, lacking economic power, are marginalized and discriminated against. Walters and Lareau (2009) show how quality educational research with academic recognition has not impacted the political world, while studies of dubious quality have served to develop legislation. This indicates that to produce political changes through research, very particular conditions are needed, which transcend the quality of educational research.

Lichtman (2011) and Ellis (2005) believe that such policies will not prevent qualitative research from developing. According to Lichtman (2011), qualitative research is currently more visible worldwide than ever before in its history. Hammersley (2008) cautions against complacency because book sales or qualitative research forums are still in demand or growing. He urges qualitative researchers to refine and strengthen the qualitative research model in light of the criticisms it has received of its internal and external validity. The following arguments are outlined to explain the exclusion of qualitative research from competing for federal funds.

> **Unfulfilled promises.** One argument is that the qualitative model did not fulfill the promises it established as a model of educational research. The qualitative model in education emerges as a response to the need to study teaching and its problems in contexts. The objective was to be able to expose the multiple processes that occur in educational scenarios and also to achieve an approach to the social phenomena that characterize the field of education. The first qualitative studies in education were based on a constructivist paradigm of phenomenological, hermeneutic, and symbolic interaction philosophies. Ethnographic, case study, and phenomenological designs clearly responded to the educational needs of understanding the social aspects of education: the student experience, the social aspect of the teaching−learning process, and the culture of the school community members. In the 1990s, this image of the qualitative model shifted to a transformative, politically charged paradigm of social critique, feminist,

postmodernist, and social advocacy philosophies of the oppressed. The oral history, feminist, social analysis, and postmodernist designs to expose the voices of the oppressed became distanced from the needs of education. The critique of these postmodernist and transformationalist studies was deadly in the field of education. The argument against these studies was that they were not scientific research but the rhetoric of abstract ideas that did not provide solutions to education problems. Postmodernist studies were caricatured, as was the case with positivist studies (Paul, 2005).

Paradigmatic fragmentation. It is argued that the diversification of research philosophies among qualitative studies fragmented the community of researchers, scholars, philosophers, and methodologists who had been constituted under the constructivist qualitative model. At the time of writing, four philosophies dominate qualitative studies: positivism, constructivism, social criticism, and pragmatism (Creswell, 2007). Among scholars and researchers, the qualitative model fragmented to losing its identity and group cohesion. Even among the so-called "qualitative researchers," practices and philosophies are openly rejected (Denzin, 2009). For example, postmodernists criticize constructivists (Mertens, 2005). This weakened the image of the qualitative model because it had no advocates. The issue of fragmentation relates to the argument that contemporary qualitative research practices evolved more as responses by individuals to their philosophical research preferences than as an organized movement to generate new knowledge and respond to the needs of education.

Distancing between constituent groups. A gap is alleged between the proponents of the qualitative model and its consumers. The idea here is twofold. One strand is that the intellectual pillars of the qualitative model have always been university academics. They have concentrated on their academic work, which distanced them from the needs of the multiple sectors of education. The other reason is that this group of intellectuals underestimated the criticisms made of the model, dismissing them as criticisms of ignorance, but these found political and decision-makers ears. The point is, although, for university academics, the field of education lives in a world of social relativities, for many administrators, politicians, and parents, the precision of quantitative data is also needed to guide and shape the generations of students with whom they work. The absence of advocacy and representation of the qualitative model in policy forums and influence excluded it from federal funding in 2000 with the Castle Amendment (2000) and the No Child Left Behind Act (2001) (Paul, 2005). Educational research has more traces of a positivist and quantitative cut in both legislations than of a qualitative nature. In this context, qualitative research can be part of federally funded research projects if used as a complementary approach to the quantitative approach or mixed methods research (Denzin, 2009). Qualitative research is now challenged to turn their findings of social phenomena into concrete evidence if they wish to compete for funding from the U.S. Department of Education.

The first paradigm war focused on the strengths and weaknesses of quantitative and qualitative research models in responding to problems in education. In the second paradigm war, the eligibility criteria for all educational research seeking federal funding are quantitative: experimental methods and probability sample studies to produce "science-based research" and "evidence-based professional practice." An open position on "science-based research" and "evidence-based professional practice" is that it does not fully apply to the field of education. The area of education is nuanced with objective and subjective situations that entail multiple research approaches. The political meddling that occurred at the beginning of the 21st century is considered a setback. For some, it returns educational research to the 1980s: too conservative values tied to the preference for experimental research methods and relegates to the background all other non-experimental research models that exist in this discipline of studies (Lichtman, 2011; Ellis, 2005). Others consider that educational research returned to the 1950s, to the beginnings of experimental research. Experimental methods have demonstrated their ineffectiveness in education due to the impossibility of controlling all the internal and external variables that affect their validity (Denzin, 2009). Many human phenomena do not manifest themselves in the social world in the same way that they manifest themselves in the world of statistical probabilities of random sample studies (Hammersley, 2007).

The Relevance of the Science-Based Research Model to Education

Denzin (2009), Elliot (2007), Hammersley (2007), Oakley (2007), and Kincheloe (2003) discuss the science-based inquiry model for eliciting evidence-based professional practice. At first glance, the science-based inquiry model seems reasonable to inform an evidence-centered pedagogical approach. Two arguments are outlined for promoting this model in education: the alleged poor quality of educational research and the alleged success it has had in medicine. From an administrative and policy perspective, to provoke improvement, comparing professions and identifying and emulating successful practices is sensible, wise, and common contemporary practice. The problem arises because the implementation of the model relies on quantitative research criteria to produce that evidence. Qualitative research suffers when compared to quantitative research criteria (Hammersley, 2007).

There are many criticisms of the NRC model of science-based research. These criticisms range from the political analysis attached to the model (Denzin, 2009) to question the methodological aspects of educational research and the concept of science that it seeks to impose (Hammersley, 2007). Denzin (2009) identifies two variations of the science-based research model. He argues that these models have proven to be ineffective in the different areas of social work that have been used, including public administration in the United States. (a) The first model is the one presented by the NRC: rigorous, objective procedures and the verification of evidence. Evidence is information that is validated through experimentation or the calculation of statistical probabilities. This model presupposes the existence of evidence that is objective, generalizable, and verifiable. (b) The second model is a simulacrum of the first model, where evidence is information that is aligned with the administrative agenda being pursued. The criteria that define evidence are whether the information appears factual, whether it aligns with the work plan, and whether it justifies the action to be taken.

Denzin (2009) criticizes science-based research models for their methodological fragility to establish the following: (a) What is evidence? (b) How is information converted to evidence? (c) Who determines what evidence is and whether it is of quality? (d) How is that evidence translated to improve practice or reality? Systematic and rigorous research methods do not translate into valid evidence. What counts as evidence is determined by the government of the day, the agency that funds the research, or the researcher who controls the study and interprets the data. The NRC model of science-based research is a one-dimensional model of doing science. (a) It emphasizes only three considerations: rigorous procedures, objectivity, and verification of evidence. It recognizes experimental and probability sample research as the best models for doing science. The model ignores needs and complex historical, contextual, and political values in evaluating educational research. (b) It excludes qualitative research because it rejects principles of this model that respond to the social complexity of educational systems: the multiple perspectives-realities, the value and cultural contexts of learning, and the functioning of educational institutions. (c) It assumes that the only way to study cause-and-effect relationships is to segment variables. This reflects insensitivity to the realities of educational systems by not considering the interpretation of the participants, their educational contexts, and their processes. To have policies sensitive to the cultural diversity of educational systems, it is necessary and desirable to have the considerable knowledge that comes from the multiple epistemologies and methodologies of existing educational research. (d) The NRC model presents a limited view of science and can be classified as pseudo-science. Hammersley (2007) indicates why educational research has to be limited to a research model that has demonstrated its limitations in education. The quantitative research model was the first research model to be used in the field of education. This model lost acceptance due to its inability to measure all the phenomena manifested in the field of education, for not producing conclusive evidence on the effectiveness of teaching and learning processes, and for not being able to produce evidence of all the objective goals pursued by educational systems. From these limitations emerges the qualitative research model that unleashes the debate of the 1980 paradigms (Denzin, 2009). (e) The research model presented by the NRC will produce only an accumulation of information about "things that work or do not work" in education through experiments based on statistical probabilities. Education is not mechanical and does not work this way (Elliot, 2007). What is paradoxical in this policy debate is that even academics and educational researchers who favor quantitative methods recognize the role of qualitative education research (Phillips, 2009).

Relevance of the Model of Science-Based Educational Practices in Education

Like the science-based research model, the "science-based educational practices" model seems logical, sensible, and wise. Criticisms of this model emanate from the instrumentalist view that imposes "evidence-based educational practice." The premises and assumptions made about education, educational research, and the conceptualization of teaching on which it is based and intended to impact are questioned. The controversy of what evidence is, which determines the quality of that evidence, and how evidence is translated into practice brings the debate back to research models. The question arises again, whether using a single quantitative research model, which has proven ineffective in education, can improve teacher education practice. To this is added a further controversy: what is the role of educational research in education under the evidence-led professional practice model?

Positions in Favor of the Evidence-Based Practice Model

The thesis of the evidence-centered or science-based educational practice movement lies in the alleged "ineffectiveness of public education" and "the poor quality of educational research," especially postmodernist qualitative research (Walters, 2009; Hargreave, 2007; Shavelson and Towne, 2002; Pring, 2000). It is argued that education is not an influential profession. Hargreaves (2007) cites the following definition from the *Journal of Evidence-Based Medicine* to describe what is meant by the evidence-based professional practice model in the field of medicine, and which is sought to be implemented in the field of education (p. 13):

> "[...] problem-based learning is a lifelong process where a patient's care, health, or other dilemmas about their care creates the need for evidence about diagnosis, prognosis, or therapy. Evidence-based medical practice is a process of converting questions about patient needs into answers, efficiently identifying the best evidence for these questions, critically evaluating the validity and usability of the evidence, and applying the results to clinical practice to assess performance."

He argues that this model will favor teaching that occurs in the classroom because education needs evidence about the educational needs of students: (a) What things work in the classroom? (b) With what populations? (c) Under what conditions? (d) What is the effect? It indicates that there are reports (e.g., Shavelson and Towne, 2002) and studies in schools (e.g., Hargreaves, 2007), where it is evident that education is not an evidence-based profession, that the quality of educational research is questionable, and that the millions of dollars invested by the government in this type of research are not justified. The problem facing education in achieving effective professional practice in the classroom lies in the gap between educational research and teaching practice.

To make education an evidence-centered profession, the current structure of educational research needs to be changed and redirected to classroom teaching (Hargreaves, 2007; Shavelson and Towne, 2002). This will involve establishing legislation to bring about changes in the structure of current educational research. One such change is to adopt a science-based research model. This will resolve philosophical and methodological disputes among researchers because there will be only one scientific research method (Shavelson and Towne, 2002). According to Hargreaves (2007), national educational research agendas must be established that respond to the educational needs of the country and the teacher in the classroom. This will entail involving professional bodies as custodians of the development of educational research. These bodies should be professional forums where educators, administrators, policy makers, and researchers express their research needs. Some researchers may understand this as losing professional autonomy because their research agendas will not prevail. This is meritorious if one thinks of the best interests of the students. Here are the arguments for making education an evidence-centered profession:

The need to change the structure of knowledge to practice teaching. The argument that education is not an evidence-based profession stems from the comparison made between medicine and teaching. When the knowledge needed to practice both professions is contrasted, differences emerge. It is argued that the scientific culture of the profession and its need for evidence for professional practice begins to be defined. Hargreaves (2007) calls this the infrastructure of professional knowledge. The field of medicine and the field of education are both professions focused on services to people. There is a marked difference between the two professions about the importance of scientific knowledge in professional practice. The knowledge that guides medical practice rests on the natural sciences: anatomy, physiology, pharmacology, etc. No physician denies the value of this knowledge for the effective exercise of their profession. The technical language centered on this knowledge is evident in practice. In teaching, the infrastructure of knowledge that guides procedure is called the foundations of education: philosophy, psychology, sociology, and history. There is no particular language for educators around this knowledge, nor a consensus on this knowledge as it applies to practice. As early as high school, the gulf between the theory and practice of teaching is evident. A teacher can be effective in his profession, ignorant of this knowledge.

The need for a scientific model of educational research. It is argued that to improve classroom practice, the gap between research and professional practice must be closed. Following are the shortcomings of academic research when compared to medical research.

> **Lack of accumulation of knowledge to guide practice.** Medical research is cumulative. Research projects explicitly seek to validate or reject evidence and theories or replace evidence with better evidence or new ideas. In the field of education, research does not proceed in this way. Much education research seeks to create new knowledge for other studies to validate or discard. There are too many small-scale studies, or studies of limited scope, on topics that other studies do not explore in depth or follow. This produces inconclusive data or data of little relevance to teaching practice.
>
> **Lack of practice in replicating research.** In the field of education, the method of replicating studies is uncommon. Replicating studies is essential in social and educational research to triangulate data by the cultural contexts of schools.
>
> **Lack of lines of research.** There are defined lines of research in medicine, some with illustrated histories and others emerging as responses to changes in society. These lines of research begin and end when the problem is solved. Researchers pursue these lines of research, and their efforts contribute sooner or later to the body of knowledge. In education, there are no defined lines of research. Research seems to follow the line of thought of the fashionable paradigm. In education, as in many social science disciplines, there are bitter disputes among researchers over the research philosophies and methodologies to be followed. Parallel to this, postmodernist approaches promote that social research is not cumulative and is another way of thinking to create knowledge. This is detrimental to educational research as it has detracted from its prestige and funding in the eyes of the policy community because it disconnects it from professional practice.
>
> **Lack of a uniform vision of what research means.** Most medical research is applied. It investigates what works and under what conditions. The goal is to achieve more accurate diagnostic methods, better patient management, or more effective treatments. The people who produce this research and publish it

are not physician-researchers but medical practitioners working in hospitals. In education, very few federally funded studies are conducted by classroom teachers.

On the contrary, most educational researchers are university professors. Often, this research has no relevance to the classroom teacher. Most studies in education, which are published in professional journals, are consumed by other educational researchers. The gap between research and practice is widened by the view that there is basic research to generate new knowledge and applied research where professional practice needs are investigated. In some universities, basic research is promoted as superior and more prestigious to applied research. In the field of education, this has been detrimental to the profession. In contrast, it has had a positive effect in medicine: physicians rely on, seek, and update their knowledge through applied research, and rely less on what they learned in their years of professional training. Another positive effect of this is seen in physicians' demand for new knowledge while teachers are not concerned with new scientific knowledge to improve their practice.

Positions Rejecting to the Evidence-Based Practice Model

The thesis of those who do not favor the evidence-based practice model is that it does not apply to education. The model is not responsive or sensitive to the role of the educator in the classroom. It changes educational research position, and its implementation will do more harm to the teaching profession than good. The evidence-based model of educational practice is a disguise to favor quantitative research.

Hammersley (2007) discusses the comparison made between "medical practice guided by the evidence generated by scientific research" and the teacher's educational management. He examines the role of scholarly research in the evidence-based practice model. He uses the work of Hargreaves (2007) as a source for discussion and recognizes that educational research can be improved by comparing it to research models in other professions. He questions the supposed success of the evidence-based practice model in the medical field, its potential benefits to the area of education, the supposed parallel between the two professions, and the alleged shortcomings of educational research. He argues that positions against the quality and usefulness of scholarly research and qualitative research can be misleading if left unchallenged. The complexity of the field of education cannot limit educational research to a function of informing practice. Research that questions the effectiveness of those practices is also needed (Phillips, 2009). The evidence-based practice model reduces the teacher's role and educational research to a technical aspect. Education is more complex management. Hammersley (2007) indicates that the instrumentalist vision to which study is reduced with the evidence-based practice model has done much harm to research in the fields of social sciences, humanities, and natural sciences because it politicizes it and reduces its role, scope, and contributions.

The following arguments are outlined against the medical model of evidence-based professional practice as applied to the field of education:

The doctor−patient relationship is different from the teacher−student relationship

Both professions work with people and care about their well-being. Physicians work in a one-on-one relationship with their patients, while teachers work with groups of students. The patient visits the physician voluntarily while most elementary and high-school students attend schools mandated by law and their parents. The doctor prescribes medication and the teacher's symbolic interactions in the form of knowledge, values, and skills (Hammesley, 2007). This fact has an evident impact on the dynamics established between the consumer of the service (patient/student) and the service provider (physician/teacher), and that affects their professional effectiveness. An example of this translates into the need for teachers to motivate and convince students of the importance of their education. Most medical research occurs in laboratories and educational research in schools and communities. Most medical research comes from the physical world, and most educational research comes from the social world of schools. Most medical researchers are practicing physicians. Most academic researchers are university professors who are former school teachers and are familiar with the educational systems they do their research. Although educational research, and education as a profession, may be enriched by comparison with medicine, it is not an accurate comparison because there are marked differences between the two disciplines.

Medical research is just as imperfect as educational research

Medical research is presented as a successful model. There are documented scandals that show serious problems with research in this profession: inappropriate designs, non-representative samples, tiny samples, incorrect methods and analyses, and erroneous interpretations. The problem is that these studies exist and some are financed with public funds, but they are not known because professional journals do not publish them.

Evidence-based practice is not accepted by all physicians

Sectors of the medical community have questioned the evidence-based medical practice model. The concept is confusing because the medical profession has always been a research-based one. The model overemphasizes the use of study results as if they were the only evidence to consider when making decisions. For example, physicians consider a person's medical history and their prior and immediate evaluations as part of diagnosing and prescribing. The explosion of generating new evidence is reflected in the quantity and quality of the medical literature available. There is a lot of information on some topics and very little on others. This creates problems because the physician must adhere to treatments and drugs where there is evidence, even if it does not satisfy their feelings or medical experience. The emphasis on the use of current evidence forces physicians to rely on the knowledge of their professional training. The literature is beyond the capacity of physicians to keep up to date. The response to this has been incorporating abstracts that allow physicians to make frequent use of this literature to keep them updated. It has become evident that the development of these summaries brings with it distortions in the information they present. The distortion comes from how it is written or the relevance of what is considered essential to include or exclude from the summary. Why is it necessary to radically change education to structures that have historically proven questionable in other professions?

Professional expertise and experience transcends research data

In medicine, there is a lot of research and information on the medications that are prescribed and on the use of protocols to be followed in dealing with medical events. The physician's analytical and interpretive capacity of each patient's circumstances determines his or her final decision on what treatment to follow and what medications to prescribe. The same is true for the teaching practice of teachers in

the classroom. The teaching management of teachers in educational systems is not reduced to a mere mechanical practice of teaching content by adopting protocols, nor will it improve simply because they adopt teaching techniques that are statistically proven to be effective in experiments. Teaching management is a complex social interrelationship between the educator, his/her students, the contents, the parents, and the other employees of the educational system. The intention is to humanize, develop values and knowledge, and produce new generations of employees, citizens, fathers, and mothers. Educational research is there to inform practice and not displace or replace teachers' analytical and critical capacity (Hammersley, 2007; Elliot, 2007; Oakley, 2007). Hammersley (2007) argues that behind the science-based research model for informing the teaching function are very different, and even limited, views and understandings of the role of the teacher in the educational process and educational research. The following assumptions and contradictions emanate from the evidence-based practice
model.

> **On teaching.** It underestimates the critical capacity of teachers because it is assumed that they work in their classrooms by intuition and not by all the knowledge that educational research has generated for years. This scientific knowledge nourishes teacher preparation programs in aspects such as the profession's role in society, learning, teaching methods, learning assessment, classroom management, cultural diversity, and human relations. It presupposes that the "science" behind evidence-based teaching practice will improve education because educators are held accountable for the learning problems of educational systems. This does not consider the structures, environments, or working conditions in which the teaching process occurs and where teachers have to work to elicit learning from their students. Although the complexity of learning is recognized, the social problems of students, their families, and the communities in which many schools are located are not considered when assigning responsibility for the poor quality of education. The interpretive element of the stakeholders' educational process is ignored in the whole equation of how the quality of education will be improved with the evidence-based practice model.

On research. It assumes that educational research is not effective in education without considering its history and contributions in learning, teaching models, curriculum, and educational administration. It is alleged that educational research is fragmented and that its contribution to practice is brief. Educators are blamed for not using research in training. Educational researchers are blamed for pursuing topics of interest when they should be researching the needs of educational systems. For this allegation, the evidence-based medical practice model is used as an example of comparison. This comparison does not consider the professional certification regulations that many educational systems apply when recruiting new teachers and where educational research is not always a hiring requirement. These regulations are requirements that emanate from political mandates. They also do not consider the range of human and social behavior issues in educational settings that impact learning. It is assumed that all education systems do not have research agendas with defined topics and needs, do not fund research, and do not have data banks to collect the information generated from research in their education systems.

Science-guided teaching practice presupposes "evidence" that is generalizable or applicable to any educational setting. Such evidence exists only in the world of statistical probabilities and not in the world of human relations in educational systems. Research to produce proof that improves only the effectiveness and efficiency of some academic disciplines' teaching and learning processes will not impact the structure of knowledge that students can develop because the purpose of education is broader. In the same way that there is a model of scientific thinking, of rational logic, there are also models of thinking in values and ethics. Educating the new generations implies humanizing them. Educational research has to link its evidence to the human aspect of its students' development and all its objectives and goals. In this way, it will have policies that improve education as indicated by the political world. Qualitative research brings that understanding to the study of human interpretation and meaning of actions in educational settings, which sometimes observation and quantitative description of human behavior do not capture.

The aim is to improve education through a research model

It is recognized that, from the perspective or structure of medical research, educational research has opportunities for improvement; but how valid are these criticisms (Hammersley, 2007; Elliot, 2007; Oakley, 2007)? There is no doubt that educational research can improve in the way it builds on previous research. It needs to validate many findings in a descriptive or explanatory way. This involves replicating studies to produce an accumulation of knowledge that supports the development of theories. Unlike medical research, in educational research, there are quantitative and qualitative research models. Each of these models has different ways of addressing the deficiencies that are attributed to scholarly research.

There is no definition of what is meant by quality educational research

The quality of educational research is criticized, but there is no indication of what criteria are used to evaluate it (Hargreaves, 2007; Shavelson and Towne, 2002). There is no definition or explanation of how the accumulation of knowledge should occur or what is considered evidence in education. Scientific research is spoken of with clear allusion to positivist research, as if non-positivist research were not equally illuminating, valid for understanding, informing, and developing classroom practice, policy development, or educational theories.

The paradigm debate: professional divisions or scientific maturity

The philosophical and methodological debates among educational researchers are categorized as unfavorable, which affects their development (Shavelson and Towne, 2002). This means a profession recognizes the problem in scholarly research to produce conclusive knowledge about social phenomena and have a science of human behavior. The differences among educational researchers are a healthy sign of the development of educational research (Okley, 2007) and the development of its scientific mindset (Hammersley, 2007). The science of educational research does not resemble research in the natural sciences. For example, from a natural science perspective, it is monumental to pinpoint the phenomena of teaching or learning styles to be investigated and measured. Measuring constructs, such as these, create controversies with the validity or reliability of measurement and replication studies.

The alleged lack of accumulation of knowledge
There is criticism that educational research has produced much-unvalidated knowledge and that it is not cumulative. Still, there is no recognition of the efforts that educational researchers have to make to respond to the political agendas that each change of government brings and how these define and redefine the problems and national needs of education. There is no recognition in the critique that educational researchers have to do research that responds to the requirements of external funding agencies, to the requirements of the educational institutions that hire them, and to the demands that the academic discipline establishes (Hammesrley, 2007).

The allegedly poor relationship between research and education policies
The poor relationship of educational research with the development of educational policies is criticized (Hargreavey, 2007), but two tendencies of the field toward educational policies are not recognized. One trend is action research, where the constituents of educational systems assume a participatory role in research and the changes that accompany it. This includes the development of educational policies. The other trend is the visible requirement of many foundations that invest funds in educational research to require researchers to inform study participants or the school community of nature and objectives being pursued. These funders require the researcher to make explicit how the study will impact classroom practice or policy development. These funders also make it clear that they are the ones who have the right to publish the findings. Some educational researchers clearly state how their work contributes to educational theories, either by expanding or questioning them. This is also not recognized; it is simply ignored.

Methodological Evolution

The second paradigm war brought to the discussion the concepts of the usefulness, scope, and relevance of educational research to solve educational problems and prescribe its professional practice. This discussed the methodological, contextual, and structural aspects of educational research that need to be addressed to improve its scientific effectiveness. In their book, *Philosophy of Educational Research in a Global Era: Challenges and Opportunities for Scientific Effectiveness*, Ponce, Pagán-Maldonado, and

Gómez Galán (2017) discuss these issues and organize them into the four main challenges facing educational research in the 21st century.

The political nature of education. The second paradigm war makes it more than evident that the field of education is one where partisan politics reigns. Partisan politics creates challenging conditions for educational researchers. (a) It entails investigating political ideals that take the form of educational standards, goals, and objectives, often ambiguous and difficult to track and measure. This implies that educational researchers have to be very creative in designing research to respond to these demands. (b) Sometimes, the research requested is a commissioned study where the desired outcome is expected. (c) Sometimes, the assignments generated contradict the work plans of policy makers, which brings problems for educational researchers.

The dislocation between educational research and practice. The discussion about the usefulness, scope, and relevance of scholarly research to improve education brought to debate the relationship of educational research with the preparation of the profession. For some, the relationship between research and practice has been controversial (Peñalva, 2014), imperfect, and, sometimes, non-existent (Scheneider, 2014; Lysenko, Abrami, Bernand, Dagenais, & Janosz, 2014). This has raised questions about the quality and usefulness of educational research.

The definition of educational research as a science. With the second paradigm war, the need to define educational research as a scientific field emerges. Educational research is a social construct that has not been determined. From its beginning, educational researchers have debated in a terrain of knowledge between natural sciences and humanistic philosophies (Gil-Cantero and Reyero, 2014), which underpinned the study of education as a university discipline (Condliffe and Schulman, 1999; Condliffe, 2000; Johannigmeier and Richardson, 2008). As a science, educational research developed in an epistemological diversity of beliefs about education (e.g., utilitarianism-humanism), diversity of methodologies and research techniques (i.e., quantitative, qualitative, and mixed research methods), and forms of knowledge about the field of education (e.g., objective-real vs. subjective-constructed). In its development, the epistemological diversity of educational research was understood among educational researchers as a constructive plurality

(Hammersley, 2007; Walters, Lareau, & Ranis, 2009) and not as scientific relativity and reflection of an imprecise science (Shavelson & Towne, 2002). For example, the epistemology of the natural sciences operates from the belief of pure objectivity and value neutrality, intending to control, predict, or manipulating the phenomena being studied. From the humanities and philosophy, education is influenced by the diversity of subjective and relative postulates about the human being, the origin of knowledge, and the possibility of knowing the social world. In the 21st century, it is evident that the roots of education in the natural and social sciences have not allowed educational research, or the field of education, to be a homogeneous community (Lee, 2010). This diversity has been scientifically ineffective because there has not been a uniform position on the nature of the knowledge generated from education and how to investigate it. Educational research needs to be redefined in the face of all the conditions of education in the 21st century (Thompson, 2012). Educational research is an overly broad and very ambiguous concept that needs to be defined and structured in practice. If this is achieved, it becomes easier to define each educational research model's roles, participation, and scientific contributions. This will help to deal with the questions of quality and usefulness that are currently being debated.

The lack of definition of the construct education as a research phenomenon. Another argument that is outlined to explain the ineffectiveness of educational research is the indefinition of the construct education as a scientific research phenomenon (Pring, 2000; Ellis, 2005; Touriñan, 2014; Ponce and Pagán-Maldonado, 2015 and 2016a). For example, what is education, or what is the best way to investigate it? Education is a phenomenon for which there is no universally accepted definition (Pring, 2000; Touriñan, 2014). Since its beginning as a profession, education has been investigated from the premise that it is a natural phenomenon or that it is a social phenomenon. These views of education emerge with the quantitative and qualitative research methods adopted in scientific research in education. The emphasis of educational research from these optics was the method and not the construct education (Cohen and Manion, 1980; Carr and Kermmis, 1985; Condliffe, 2000; Shavelson and Towne, 2002; Johannigmeier and Richardson, 2008; Ponce and Pagán-Maldonado, 2015; Ponce, 2016). The result of this is the absence of a shared vision of education as a scientific research phenomenon (Ponce and Pagán-Maldonado, 2016a). Educational research in the 21st century is how the field of education is defined (Pring, 2000; Lee, 2010). The problems of the effectiveness of educational research have to do with the nature of the data, their origin, and how they emerge (Koichiro, 2013). For this, educational research needs to empirically construct education as a research phenomenon (Touriñan, 2014) and develop a universal language to capture educational reality (Thompson, 2012; Smeyers, 2013). This topic is discussed in Chapter 7.

The discussion of the utility, scope, and effectiveness of educational research in solving academic problems and prescribing its practice sparked other controversies with scholarly research that transcend the focus of this book. Ponce, Pagán-Maldonado, and Gómez Galán (2018) discuss these issues in their book *Educational Research Issue in a Global Era: New Frontiers*. The discussion of the usefulness, scope, and effectiveness of educational research sparks interest in scientifically investigating the construct of education. This is the subject of Chapter 7.

7

Scientific Research and Education

José Gómez Galán

The problems of education lead to scientific research on education. Educational research eventually turns education into an object of scientific investigation. For much of the 20th century, education was studied from philosophical and semantic conceptualizations. At the time of writing this chapter, there is no universally accepted view of what education means. Educational research focused more on solving academic problems than on understanding education as an object of scientific investigation. This approach proved to be of little scope in the face of the complexity and challenges that education presents in educational systems. In this beginning of the 21st century, education begins to be studied scientifically through the activities that operationalize it (Ponce, Pagán-Maldonado and Gómez Galán, 2017 and 2018). This chapter examines education as an object of scientific research.

Education as an Object of Scientific Research

Education is an ideal that has developed over time. This is known as a social construct. Education is understood as the acquisition of knowledge, values, skills, and attitudes. These allow the student to build or rebuild himself or herself to become a better person or become an employee. Education is an internal activity of the person in which he/she is allowed to think, feel, want, operate, project, and create. In this way, education is externalized because it enables the person to play, work, exercise a profession, or respond to situations. Educating implies that each constituent of the educational process assumes its responsibility and commitment to education (Touriñan, 2014). The visions, definitions, and conceptual positionings that exist of education are helpful because they allow us to separate what is known from what is not known conceptually. The problem with abstract definitions is that they do not always capture the actual activities that shape the concept or the social construct called education; hence, the need to scientifically investigate the construct of education.

As an object of scientific research, education is just beginning to be studied through the activities that make it operational. The argument is that scientifically investigating the construct education provides for understanding the theory and practice of education to generate the theory that explains education. Explaining education is essential to improve it. Scientific research on education is necessary to know education as a realm of reality rather than a mere philosophical postulate. This knowledge should facilitate the ability to describe, explain, understand, interpret, and transform education from its methodologies, concept and constructs (Touriñán, 2014).

A scientific approach to the construct of education comes from the study of its educational practices. Studying educational activities allows us to understand their features. The educational activity intends to educate and conform to the intention of the criteria of the action of educating. The training entails giving it character, meaning, determination, and qualification because they construct education. An educational activity demonstrates its logical necessity of use, has a theoretical and practical reason of principles, and calibrates with the intention of the practice it pursues.

Educational activities are means of intervention to achieve education. It implies advancing between the cultural knowledge needed by the student, the pedagogical intention pursued, and the usefulness of learning or its application. Every educational activity starts from the pedagogical vision of the educator. It is for this reason that in education, there are methods of thought and methods of action. Education problems lie in the link between knowledge, effort, and the contextual intention pursued in objectives, goals, or standards. Education in educational systems is standardized through their educational policies. These are standards of assessment on how to do the educational practice. Without knowledge of education, there can be no standards, and without means, there is no valuation of education (Touriñán, 2014).

The quality of teaching is considered the most crucial criterion in the quality of education. Years of research have not yet produced a uniform teaching standard applicable to all contexts to make quality education. In the literature, teaching grade is a concept without uniformity and consensus. There are too many variables that interact when explaining the quality of teaching. There is no consensus when it comes to attributing weight to them or explaining a causal relationship. For example, some link the values, beliefs, and knowledge that teachers bring to the profession with the quality of teaching. Standardized tests can measure these knowledge, values, and ideas.

Consequently, teaching quality is associated with prospective teachers' performance on professional tests or revalidations (cognitive resources). A second variable lies in the credentials of teachers and their certifications (credentials). A certified teacher implies that he or she meets quality requirements to practice as an educator. A third variable is whether the teacher has the knowledge and skills commensurate with the quality standards of the educational systems. Quality means that the teacher enters the education system with professional training that is consistent with the standards. Once in service, there is training to ensure that educators maintain the knowledge to practice. A fourth variable is that the quality of teaching lies in the activities that the teacher does in the classroom to produce learning. Teaching scientifically means that the educator uses educational activities that are scientifically based or scientifically proven to be effective. Finally, it is argued that the quality of teaching lies in the quality and quantity of teachers' experiences in and out of the classroom. This explanation emphasizes the training of teacher preparation programs and the lifestyles of educators (Wang, Lin, Spalding, Klecka, & Odell, 2011).

The product of education is learning (Ponce, 2014b; Carneiro, 2015; Ponce, Gómez Galán, & Pagán-Maldonado, 2018). Learning is conceived as changes in students. Learning is studied as an individual student phenomenon and an institutional product (Ponce, 2014b and 2016; Ponce and Pagán-Maldonado, 2016a). Education in the 21st century is more about learning than education: learning to know, learning to do, learning to live and work, learning to be, and learning to learn (Lee, 2010; Carneiro, 2015). Education in the 21st century seeks an inclusive world (Carneiro, 2015) in a global culture of economic competitiveness (Colella and Diaz, 2015). In the 21st century, education is a much broader concept than what happens in educational institutions.

In the 21st century, learning is determined based on standardized test results for K-12 public education systems, learning assessment mechanisms in North American universities, and educational research-evaluations in the educational systems of many European and South American countries (Ponce, 2014b; March, 2015; Phillips, 2015). According to LeCompte and Aguilera-Black Bear (2012), assessing learning simply means asking whether institutional learning goals and objectives were achieved, under what criteria or standards they were performed, and how well they were achieved. Joint exercises to complete practical assessment are to identify and define the learning competencies to be assessed at the institution, align curricula to identify the courses where this learning is being addressed, set clear learning objectives to achieve precision in assessing achievement, and establish consensus on how to evaluate learning. This process is relatively free of problems or controversy if institutional agreements are in place. The dispute with education arises when data must be compared across institutions for accountability purposes. The absence of a common language of learning and appraisal then arises (Gómez Galán, 2016; Ponce, Pagán-Maldonado, & Gómez Galán, 2017 & 2018; Ponce, Gómez Galán, & Pagán-Maldonado, 2018). For example, although there are codes of course validations among universities, it is impossible to pinpoint whether the educational experience in one private university's Ethics 101 course is the same as another's Ethics 101 course. According to LeCompte and Aguilera-Black Bear (2012), there is little scientific research to support the range of quality standards being applied to colleges and universities by regional accrediting agencies or professional bodies. Although the assessment models show their scientific rigor, the validity and reliability of the data change according to the political standard with which they are evaluated. For example, in the United States of America, after decades of documenting how universities fulfill their institutional missions, there is still much uncertainty about educational quality (Palmer, 2012; Ponce, 2014b; Smith, 2015). In the United States, the accountability movement has worked on standard criteria and indicators of learning and has created national data banks to facilitate comparison between institutions. However, standardizing education remains a challenge (Palmer, 2012).

The Complexity and Dynamism of Education as an Object of Scientific Research

Educational research is carried out in specific educational contexts of schools and educational systems to observe education. Education, as an object of scientific research, manifests itself as a complex and non-static phenomenon due to its multidimensional dynamics of symbolic interaction.

Multidimensional Phenomena

Education as an educational system generates "multidimensional" phenomena. A multidimensional phenomenon means that its origin may be due to many factors or causes. For example, student learning is not only the result of teachers' teaching strategies, but student motivation, peer relationships, or parental supervision also plays a role.

Symbolic Interaction Phenomena

Many educational phenomena are symbolic. This means that they can be studied from different perspectives and all produce diverse and valid data. For example, learning can be learned from the teachers who work it, from the point of view of the students who receive it, or from the point of view of the administration that supervises it. This premise has generated controversy among educational researchers and academics. Education causes phenomena of symbolic interaction, and for others, it does not (Ponce, Pagán-Maldonado and Gómez Galán, 2017 and 2018).

Dynamic and Non-Static Phenomena

Education produces phenomena that are not static but dynamic and evolutionary. This means that it is difficult to identify its beginning, its development, and its conclusion. For example, learning does not always progress linearly. On the contrary, a student may finish one academic year with excellent grades, only to return the following year after summer vacation and forget everything he or she has learned. Another example is student dropout. A student may drop out of school and surprise his teachers and parents because there was never a straightforward or observable indicator that culminated in dropping out.

Challenges for Educational Research

The complexity and dynamism of educational phenomena have confronted educational researchers with three challenges when studying education.

The Challenge of Conducting Contextual Research

Educational researchers have to conduct their studies with students who are influenced by the environment of their schools. This is known as contextual research. For example, a sick child, the complex divorce of a student's parents, an abused child, or a new school principal. This affects educational research settings because the study's contexts cannot be controlled or changed

(Berliner, 2002). It is often difficult to pinpoint the influences of educational contexts on student behaviors. Each school has its teachers, students, and staff. This explains differences in programs but not differences between programs. It is for this reason that many educational reforms fail when implemented in schools other than those studied. In educational research, context is essential because it allows us to see profound aspects of educational scenarios and the knowledge that exists in them (Peters, 2012). To respond to the expectations of educational improvement and the development of educational policies, educational researchers have to conduct their studies as close as possible to the reality and circumstances in which the phenomena they study are manifested to observe them, describe them, and measure their effects on education or institutional effectiveness (Mejías, 2008; Ravitch, 2014).

The Challenge of Researching Socially Complex Scenarios

Another complexity with contextual education research emanates from the institutional structures that characterize education systems. For example, many educational systems have a three- and four-tier functional design. A first-level tends to be administrative offices that oversee various schools, districts, or educational regions. The second level of operation is administering each academic unit such as a school or an academic department in a university. The third level of organization occurs in the classroom. A fourth level is extended after-hours programs to meet the needs of students, adults, or the community. There are educational projects whose operation involves all levels, and this makes the work of educational researchers more complex when studying educational phenomena and determining the effectiveness of programs. For example, the assessment of learning is done by the faculty with the students. Still, its policies may originate from a regulatory agency or high management level and be supervised by an intermediate level. Its implementation is done by faculty, and its data can be used for decision-making at all levels. Research-evaluating, the effectiveness of a learning assessment program, may involve using different research strategies and applying various educational stakeholders. Educational systems are complex because they involve the interests of many people (government, administrators, educators, students, and parents), work with diverse values, and seek to channel these toward educational goals and objectives. When studying multidimensional phenomena in the complexity of educational contexts, researchers have two options: they simplify their approaches to explore only some aspects of educational phenomena or address these to their full extent and complexity (Ponce, 2016; Ponce and Pagán-Maldonado, 2016a). In both cases, educational researchers have to accept and manage the complexity and ambivalence of education when researching it (Berliner, 2002). If the study is simplified to be carried out, it may result in a survey without scope to contribute to the improvement of education. If the study tries to address the full complexity of education phenomena, the research may become unmanageable (Ponce, Pagán-Maldonado, & Gómez Galán, 2017, 2018). These challenges are visible when educational researchers have to identify the research design that best responds to the research objectives or explain the validity of their data or the scope or generalization potential of their study (Ponce, 2016).

The Challenge of Researching the Context of Educational Institutions

Education is such a complex phenomenon that there is also a need to develop educational research away from the classroom where education products are observed. Educational research must go outside the educational contexts where education occurs to understand other relationships about how education occurs (Lee, 2010).

Methodological Evolution

Researching education is complicated by the complexity of educational phenomena. A central element of the quantitative versus qualitative paradigm war focused on the strengths and shortcomings of these models in capturing the complexity of education. The emergence and acceptance of mixed methods research in educational research lie in its potential to capture the complexity of the profession (Ponce & Pagán-Maldonado, 2015). The criticism of mixed methods research is its complexity to be implemented by individual researchers. In other words, mixed methods research is feasible when implemented in teams of researchers. Capturing the complexity of education represents the major methodological challenge facing educational research. As will be discussed in Chapter 8, the complexity of education as a scientific research phenomenon has led educational researchers to propose the need to adopt or develop a model of educational research that is responsive to the dynamic phenomena of education. The feeling of these researchers is that current research models do not respond to or capture the complexity or dynamism of education (Mejías, 2008; Lee, 2010; Clark, 2011; Thompson, 2012; Rudolph, 2014; Gutiérrez and Penuel, 2014; Snow, 2015; Coburn and Penuel, 2016; Glass, 2016; Ponce, Pagán-Maldonado, and Gómez-Galán, 2017 and 2018).

8

From Research Applied to Practice to Research Embedded in Practice

Omar A. Ponce

The complexity and dynamism of education as an object of scientific research has raised questions among some scholars and educational researchers about the scope and effectiveness of current research methods in education. The questioning lies in whether contemporary research methods respond to the phenomena being investigated in education. The question is not whether research methods should be quantitative, qualitative, or mixed, as was the case in the 20th century, but to what extent these methods allow capturing the complexity and dynamism of the phenomena of education and generating valid and generalizable data needed to guide the practice and development of educational policies (Mejías, 2008; Lee, 2010; Clark, 2011; Thompson, 2012; Snow, 2015; Ponce, Pagán-Maldonado & Gómez-Galán, 2017 & 2018). The central idea in this line of thought is that educational research methods should respond to the particularities of educational phenomena (Rudolph, 2014; Gutiérrez and Penuel, 2014; Coburn and Penuel, 2016; Glass, 2016: Ponce, Pagán-Maldonado and Gómez Galán, 2017 and 2018). This chapter discusses the arguments about the need for a new model of educational research.

The Need for a New Model of Educational Research

In the scientific culture of the 21st century, it is recognized that science is an activity that occurs in a diversity of academic communities with their methods, standards, and practices that are highly contextual to their respective disciplines. It is for this reason that new research methods emerge from time to time. For this reason, research methods cannot be understood as separate from the phenomena being investigated. After all, science is about understanding what works and why (Rudolph, 2014). The research methods used in education come from psychology, sociology, and social sciences and have been adapted to address the phenomena of education (Gutiérrez and Penuel, 2014; Phillips, 2014; Rudolph, 2014; McDonnell, 2016; Glass,2016; Ponce, 2016; Ponce, Pagán-Maldonado and Gómez Galán, 2017 and 2018). Educational research methodology needs to be flexible, be sensitive to the cultural context, and be evolutionary to track and understand how students and teachers adapt educational practices to the institutional realities they live (Gutiérrez and Penuel, 2014). Educational research will advance when a research model for education that responds to the phenomena of the profession is adopted (Rudolph, 2014; Snow, 2015).

It is argued that education is not a static research phenomenon. On the contrary, education is a dynamic phenomenon and responsive to the institutional context in which it occurs. It is necessary to understand students who sometimes do not cooperate, sometimes are not interested, and sometimes do not even follow instructions. Classrooms are scenarios where multiple relationships and political and cultural influences are observed. Educational phenomena have ill-defined variables that are not constant. Education as a research phenomenon is very different from physical phenomena studied in laboratories (Rudolph, 2014).

Educational research is a domain where there is a tremendous epistemological and methodological difficulty. For example, its practice is characterized by being liberating at times or by economic limitations at others, by governmental interference without warning, or by methodological impositions at times. The element that has contributed most to the effectiveness of physics research is the ability to predict. A hypothesis is put to the test, and with the accurate prediction, it is set to experimental verification or refutation. In this way, truth is retained, and error is eliminated. In the field of education, this form of progress is rare and sometimes impossible. Generalizing the findings of a study is incredibly complex. It cannot be satisfactorily addressed in how it has been done so far because very little education research has the quality to make predictions, as is the case with laboratory studies. There are excellent quality studies that deeply describe and produce an understanding of educational contexts. All their elements are considered and illuminate the same phenomenon as it manifests itself in other settings and cultures. This is a prediction. In education, it is not easy to control all the variables that come into play, which are essential in explaining learning. The aspiration in these cases is to achieve ecological validity and not a statistical prediction. Educational research is complex because learning is a phenomenon that involves people in complicated and specific contexts that cannot be extracted from the context to be studied, where subjects have gender, sexual orientation, socioeconomic status, ethnicity, culture, interests, and things that bore them, where some eat breakfast and others do not, live in neighborhoods where there are shootings. Others do not (Phillips, 2014).

Educational research has mainly been *ex post facto*, where it draws on the school community members' experiences, perceptions, and experiences (Clark, 2011). Many problems in education have to do with budget, physical structures of schools, school management, and issues in educational institutions that are not necessarily subjects of research (McDonnell, 2016). Scientific research cannot solve all problems in education because issues are not solved with the knowledge of teachers' experience (McDonnell, 2016).

In the 21st century, the central question is what direction the field of educational research should take in the face of the complexity of education and its need to generate reliable and generalizable data. The answer to this question, and what separates educational research from other forms of research, is its influence on educational systems' practice and educational policies or the link between research and the preparation of the profession. Educational research has always been interested in education; so the relevance of research to practice is not a recurring theme in educational research. Of course, renowned researchers have claimed that the link between educational research and practice is almost non-existent and have recommended that educational research focus on generating theories in the face of the little effect that applied research has had because of the challenges of implementation, contextualization, and group diversity in the field of education (Rudolph, 2014; Snow, 2015).

In the 21st century, educational research must focus on studying the needs and problems educators have in their classrooms and schools. These needs need to be captured in the form of research agendas and manifestos. Collaborative links need to be established with school systems to identify successful classrooms and study them. Constant evaluation of school problems is also required to keep educational research relevant. This brings problems for researchers and educators, but if academic researchers look at what is appropriate, the possibility of impacting education is greater (Snow, 2016).

From Research Applied to Practice to Research Inserted in the Practice

Given the complexity and dynamism of education, it is argued that educational research needs to identify its research model that will help it organize and define itself as a unique field of scientific research (Gil Cantero & Reyero, 2014; Ponce & Pagán-Maldonado, 2016; Ponce, 2016 & 2017; Ponce, Pagán-Maldonado & Gómez-Galán, 2017 & 2018; Ponce & Pagán-Maldonado, 2017; Ponce, Gómez Galán & Pagán-Maldonado, 2017). Educational research needs a theory that allows it to generate scientific knowledge about educational practices and their policies. The role of theory in education is to understand and explain. The application of any approach to an empirical instance involves reflecting on the event and the context being studied to understand, help construct and explain the object of research, and generate or construct knowledge. The role of theory in research is imperative in the emerging era of globalized educational policies where the aim is to summarize students through standardized tests. The use of ideas to address social phenomena is essential because they have to be studied in the disciplines that seek to work with them. In the case of education, this means that a phenomenon cannot be described in isolation from the other social factors that surround it and affect it in one way or another. Theories should help understand the inside and outside of a social phenomenon and link it to its past, present, and future. Approaches should help to describe, explain, or emancipate. Researching education entails dealing with the complicated, confusing, impure, and uncertain phenomena that manifest themselves in this profession. This may involve rejecting the prevailing culture of philosophies and methodologies that dominate the field. In an era of accountability, numbers and measurements are the only means of generating data regardless of how the standardization of tests and curricula infringes on students' individuality (Luke, 2011; Ponce, 2018). Education cannot be studied solely by looking at the individual or its students. It is necessary to check students about other students and components of teaching and society (Lingard, 2015). Educational research needs a more complex look at education, avoid simplification, and better understand the processes and contexts of education. At this beginning of the 21st century, several authors point to complexity theory and the philosophy of critical realism as the platforms for organizing educational research and addressing the complexity of education (Cochran-Smith, Ell, Grudnoff, Ludlow, Haigh, & Hill, 2014; Gil Cantero & Reyero, 2014; Ponce, Pagán-Maldonado, & Gómez-Galán, 2017; Ponce & Pagán-Maldonado, 2017; Ponce, Gómez Galán, & Pagán-Maldonado, 2017).

In the 21st century, educational research needs to transition from the "applied-to-practice research model" or applying science to the practice of education to one of "practice-connected or practice-embedded research" to avoid the gap that exists or is created in the way education research is conducted. Interest in this research model began to develop. It was realized that much of the applied research in fields such as anthropology or child development did not translate into adequate knowledge in education because the profession does not have analogs to applied research. Much of the lack of analogy emerges from educational contexts. Topics such as reading comprehension are examples of how there are topics in education that do not conform to basic or applied research. This territory is called research embedded in practice.

Another problem with the applied-to-practice research model emerges from the unchallenged premise of how knowledge from a basic or applied study applies to educational practice. It is assumed that the insights from these studies involve using the information in educational processes. An example of this was the teaching of a second language using cross-linguistic contrasts analysis. Throughout the history of the American Educational Research Association, several presidents have called for educational research that responds to the needs of educators. Progress in education will come from the use of multiple approaches that generate instruments, such as curricula and programs, that serve educators and researchers' needs and generate reliable information. This is possible if educational research adopts a "research embedded in practice" (Snow, 2015). At this beginning of the 21st century, "design-based research (DBR)," or an educational research methodology that attempts to connect research with educational practice, emerges. The objective is to increase the transfer of knowledge between research and practice. Educational research should allow the generation of new knowledge and validate it in practice. The following characteristics are present in DBR that would allow producing educational research embedded in practice to simultaneously generate new knowledge and validate it (Anderson and Shattuck, 2012):

1. **Being in the educational scenarios**. Being in the academic context increases the validity of the knowledge generated, at least in the background where the research is developed.
2. **Focus on the design to test an intervention.** An effective strategy is to migrate effectively from the experimental classroom to regular classrooms, with regular students and regular educators. Researchers and educators jointly design the instructional intervention with its theory and implementation sequence.
3. **Mixed methods.** Researchers need to be pragmatic in their approach to study all the dynamics that develop in implementing the intervention. This implies the use of several research methods in the same study.

4. **Improvement in implementation.** The intervention is improved in practice to obtain positive results.
5. **Collaboration between researchers, educators, and students.** Understanding the scope of the intervention involves the cooperation of all actors to understand the multidimensionality of education.
6. **Evolving design and action research.** The study design will grow as the intervention progresses and develops to capture the dynamics that emerge. The study is modified as it goes along.
7. **Practical in implementation.** Procedural and performance issues are eliminated because the researchers are embedded in the study. This should produce quantitative data without reference to the margin of error because there are no probability samples. It also creates a detailed description of the processes and attitudes of the students and educators which allows the results to be appreciated.

As a methodology, the use of embedded research-in-practice model seems to be observed more in the United States of America. The method has yielded positive project and school improvement results. It has been used in phases of years where various researchers have intervened to study complete cycles of conceptualization, implementation, evaluation, and follow-up interventions.

Methodological Evolution

The practice-embedded research model responds to four current gaps in educational research: (a) it generates knowledge from practice, (b) it validates new knowledge in practice, (c) it increases the generalization of knowledge about practice, and (d) it responds to the needs of educators, students, and the realities of educational systems. The potential of the model of research embedded in educational practice is obvious. Is research embedded in practice a change of vision and philosophy of research or a new model of educational research?

The implementation of practice-integrated research constitutes a departure from the current scientific culture of educational research. For example, the practice-integrated research model involves close collaboration between educators and researchers. This may include the presence of researchers in classrooms, whether physical or electronic. This, in turn, may entail a change in the dynamics of the groups in their classrooms. Operationalizing the research model embedded in practice could involve compliance committees making amendments to subject protection protocols to make this research approach feasible. It could also entail collaborative agreements between schools and universities for researchers to have access to classrooms. Another departure from the current scientific culture of educational research is the time variable or the duration of the study. Research embedded in practice may require more time to observe the phenomenon, track its manifestation, and understand its impacts in the context of the school organization that governs the school.

In summary, the practice-integrated research model advances the state of the art of contemporary educational research because it provides answers to existing gaps. Its operationalization merits much more discussion. This discussion need not trigger a third paradigm war.

9

Educational Research in an Era of Ethics in Scientific Research

Omar A. Ponce and Nellie Pagán-Maldonado

A Brief Historical Background of Ethics in the 21st-Century Scientific Research

Promoting responsible conduct in human and animal research has been at the core of the applied research ethics movement. The institutionalization of this movement can be seen in the official constitution of Institutional Ethics Committees. The functioning of these committees focuses on training in the ethics of scientific research, certifying researchers on ethics, and providing guidelines for inserting ethics in research and approval procedures to ensure ethics when conducting scientific research (Horner and Minifie, 2011; Garcia Rupaya, 2012; Paoleti, 2014; Abebe and Bessell, 2014; Guerriero and Castaño-Pineda, 2015). A common element in all these guidelines is the predominant role they give to research ethics committees, as an instance independent of researchers, which evaluates, approves, and monitors research to ensure that it is conducted under ethical criteria. The intention is that the honest evaluation of scientific research should not be exclusive to researchers but should be extended to a committee with institutional backing to provide a more excellent guarantee before, during, and after the execution of the projects. The life cycle of a research project starts with elaborating a proposal that must include an analysis of its ethical aspects. This document is submitted to an Institutional Ethics Committee, which will carry out an evaluation and then decide whether it is approved or not. Once approved, it is implemented, in the best of cases, under the supervision or accompaniment of the Committee. It presents a report where the different ethical aspects of the conduct of the study are accounted for. The publications and strategies for disseminating the new knowledge are presented (Arias-Valencia1 and Peñaranda, 2015).

Ethics in Educational Research

At the beginning of the 21st century, few question the importance of ethics in educational research (Hansson, 2011). Ethics in educational research is a theme that emerges in the first and second quantitative versus qualitative paradigm wars discussed in Chapters 4 and 6. The fact is that the diversity of philosophies in educational research has generated a wide range of methodologies on how to research education. As discussed in Chapters 1, 3, 5, and 8, educational research uses students, sometimes parents of students, educators, and administrators, to gather information for research. The nature and dynamism of education as a scientific research phenomenon, which we discussed in Chapter 7, brings the need for this methodological diversity and the need to research in and out of educational institutions to appreciate the effect of education on students and society. The issue of ethics in educational research can be seen from two points of view: (a) the need to have educational research with sensitivity and respect for the research participants because of the legality that this entails for the institutions that endorse and fund it (Davis and Holcombe, 2010); (b) the political chore of controlling scientific research and the money that accompanies this activity when it is institutionalized (Denzin, 2009). For these reasons, ethics in educational research has been gaining acceptance among educational researchers (Ponce, Pagán-Maldonado, & Gómez Galán, 2018).

Studying people is a joint exercise in educational research. After all, the field of education is about people. All scientific research involving people carries ethical considerations (Ponce, Pagán-Maldonado, & Gómez Galán, 2018). Ethics in educational research tends to relate to two aspects: morality related to the rightness or wrongness of situations or events that may arise in the research process, and to the ethics of the researcher's behavior and decision-making to handle any situation or dynamic that emerges in this endeavor (Pring, 2000). Ethics in educational research can be observed in the following aspects: (a) in how access to educational systems is negotiated, (b) in how established institutional norms and policies are handled, (c) in how the researcher relates to the study participants, (d) in how their welfare is protected and their safety is ensured, and (e) in how faithful the researcher is to what is studied when analyzing the data and writing the research report (Stutchburya and Foxb, 2009).

Ethics in educational research is a complicated topic because of its relativity in some instances. For example, the fundamental principle of the contemporary movement on ethics in scientific research is respect for the study participant. In the study of certain Aboriginal groups in Australia, it has been found that the mere fact of asking them questions is seen as disrespectful for Aboriginal people. If the researcher negotiates with them before the study, the possibility of asking them questions, and, therefore, the feeling of disrespect they may have, is eliminated. This example illustrates the relativity of ethics according to the cultural context and the importance of negotiating how the study will be conducted or how the research process will be before the research (Davis and Holcombe, 2010). According to Stutchburya and Foxb (2009), ethics in educational research is procedural and becomes easier to observe and understand in the development of the study. These are the instances where ethical issues in educational research can commonly emerge and be observed.

Access to study participants. Educational research revolves around student learning, how students are taught, and how educational institutions are managed. Typical participants in educational research are students, their teachers, parents, and educational administrators. Educational research takes place in schools and students' communities. Gaining access to educational institutions provides researchers with the privilege of exposure to educational policies, institutional norms, the culture of communities, and the values, beliefs, and ways of thinking of the institutions and their people. Gaining access to institutions and their people rests mainly on the "trust" that is placed in the scientific goals pursued by the researcher. Much of this trust emerges from the researcher's honesty, from the guarantees of confidentiality that he or she establishes so that the institution and the participants of the study do not fear to participate, to his or her compliance with the rules and conditions set by the institution or the participants to participate in the study, and to the good relationships established by the researcher in this process (Ponce, 2016; Ponce, Pagán-Maldonado and Claudio-Campos, 2017).

Interrelation of the researcher with the participants. In educational research, data are collected through field observations, personal interviews, focus groups, surveys, the creation of experiments, or historical studies (Ponce, 2016; Ponce & Pagán-Maldonado, 2016; Ponce & Pagán-Maldonado, 2017). These forms of research and data collection entail interacting with study participants in educational settings or their communities. Much of the cooperation, openness, and honesty of the participants with the researcher emerges from the trust that is generated between them: (a) the researcher's honesty and transparency, (b) his or her respect for the cultural diversity of the participants, (c) respect for the participants' free will if they wish to leave the study at any time, and (d) guarantees of privacy and confidentiality of the information they provide (Ponce, 2016; Ponce, Pagán-Maldonado, & Claudio-Campos, 2017). The most vulnerable population in educational research is children. Children have different competencies and perceptions than adults because of their age and cultural background. In research, children can become victims of exploitation if the adult is not responsible for them in the power relationship between the researcher and the study participants. The ethical implication of this possibility is the mediation of an adult or institutional entity responsible and committed to the child's safety, a relay of consent to participate, and institutional rules that regulate these dynamics and ensure ethics in research (Boddy, 2014).

The research report. When educational research focuses on values, groups, or individuals, a significant vulnerability emerges. Educational research becomes unethical because it does not make an authentic and faithful representation of the values, realities, people, or groups studied. If the trust generated by the researcher in the participants is used, and his study does not identify them or does not faithfully represent them, then the issue is ethical. There is no universal or infallible code to prevent this vulnerability of educational research (Pendlebury & Eslin, 2001; Ponce, Pagán-Maldonado, & Claudio-Campos, 2017).

The ethical dynamics indicated above occur in the development of the study and emerge with the situations where participants may react in different ways. These three points bring to the discussion the importance of ethics in research and the responsible conduct of the researcher. Some authors indicate that ethics in educational research is a matter of practicality (Pring, 2000; Ponce, Pagán-Maldonado and Claudio-Campos, 2017), good manners, and courtesy (Davis and Holcombe, 2010). At a more intellectual level, the researcher can work through these unavoidable dynamics of research if he or she critically reflects on his or her study before conducting it to identify tools to manage them.

Two strategies are recommended to evaluate the ethical dynamics of educational research. (a) **Moral theory**. Moral situations can be managed by analyzing the positive and negative consequences they may entail. The researcher should look for dynamics that carry more positive than negative consequences (Stutchburya and Foxb, 2009). (b) **The Seedhouse model of ethics**. This model involves the researcher analyzing the ethical situation that emerges and reflecting on it and its consequences to make the best decision. The model suggests that four aspects be considered: (1) the **external aspect** or whether institutional norms and policies are violated, (2) the **consequence aspect** or what implication the action carries for society, specific groups, or the study participant, (3) the **deontological aspect** or questioning how things are being done or whether things are being done correctly, and (4) the **internal aspect of the person** or whether the person feels comfortable or good about the action. The model is applied to three research steps: recruitment of study participants, fieldwork, and preparation of the research report (Stutchburya and Foxb, 2009).

Methodological Evolution

Institutional ethics committees have received a mixed reception in the United States of America (Denzin, 2009), Colombia (Arias-Valencia and Peñaranda, 2015), Peru (Garcia Rupaya, 2012), and Brazil (Beltrão de Lucena Córdula, 2015; Guerriero and Castaño-Pineda, 2015). The positive aspect in favor of these institutional ethics committees is the professional development of the members who participate. The training they offer in the ethics of scientific research turns out to be a professional gain that their members value (García Rupaya, 2012). Negative criticisms point to the question of whether institutional ethics committees have an impact on the scientific effectiveness aspired through research. Two criticisms of institutional ethics committees emanate from different countries due to their impact on educational research: (a) the bureaucracy that the researchers have to go through to obtain approval and be able to carry out the study under institutional norms, and (b) the conflicts generated by imposing an ethical-biomedical vision on educational research.

The Bureaucratization of Research

Ethics can be a complicated and relative topic (Stutchburya and Foxb, 2009; Davis and Holcombe, 2010; Hanson, 2011; Guerriero and

Castaño-Pineda, 2015). How does one restrict the line that determines a moral or immoral, ethical or unethical act? This fact is a challenge in scientific research (Hansson, 2011). The contemporary approach to the discussion of research ethics in Institutional Ethics Committees is to discuss the studies that are proposed and determine their possible ethical implications in terms of the guidelines, protocols, and international guidelines established for the operation of these committees. Rationally, the ethical dilemmas that may arise from this practice should be resolved in this way. In practice, the reality is different. In general terms, the honest assessment and monitoring of research have become bureaucratized. Both for the researcher and the ethics committees, the exercise consists of a procedure having compliance with the requirements established by rules and checklists, which is what research ethics codes have been reduced to in practice (Arias-Valencia and Peñaranda, 2015). The evaluation of multiple institutional ethics committees is also considered bureaucratic when the study is conducted by investigators from different institutions or involves entering several institutions. In these cases, researchers have to seek approval from the ethics committees of each institution and sometimes negotiate the various ethical interpretations of the same study creatively to obtain permission, enter the institution, and conduct the study (Guerriero and Castaño-Pineda, 2015). In our experience, in master's and doctoral programs where the final graduation requirement is to complete a thesis or dissertation, the process of submitting, evaluating, and approving the study makes it almost impossible for a study, no matter how simple, to take more than one academic semester. This sometimes results in prolonging the student's program of study and its costs.

The Imposition of a Biomedical Ethics Approach to Educational Research

Institutional ethics committees are based on a positivist definition of scientific research emanating from the natural sciences. This biomedical ethics model does not apply to other disciplines or forms of research in health (Arias-Valencia and Peñaranda, 2015) or the social sciences (Paoleti, 2014), and especially to qualitative research (Denzin, 2009; Tuck and Guishard, 2013; Beltrão de Lucena Córdula, 2015). In the positivist vision of research, the reality is conceptualized as an external and independent configuration to the researcher; the scientist assumes the role of neutral actor, projecting himself as exempt from the problem under investigation, and, disinterested in social transformation, he limits himself to collecting data and analyzing them and producing knowledge, avoiding that this process has an impact on the situation under investigation. On the contrary, from a qualitative, critical, constructivist, and participatory research paradigm, the role of the researcher is different. In qualitative research, it is understood that the research process requires interaction between the subject and the field since both the data and the attribution of meaning are generated in the research process as a dialogical construction. Likewise, the former is responsible for producing knowledge that guides praxis and social transformation, taking notions of social justice, most of the time explicitly (Guerriero and Castaño-Pineda, 2015). In the social sciences and education, it is evident that both research models require different interactions between the researcher and the participants of the study, that both models imply other procedures when implementing the study, and that both models constitute different ways of generating data, different ways of negotiating access to participants and of dealing confidentiality and anonymity of participants (Denzin, 2009; Giardina and Denzin, 2013).

As a consequence, research in the social sciences and education suffers more criticism and is delayed longer in its implementation than research in the natural sciences and medical sciences. Disputes between researchers and members of institutional ethics committees who reject or demand changes in social and educational research procedures for ethical reasons, even against the researcher(s)' intention, are evident in the literature (Denzin, 2009; Parsella, Amblerb, & Jacenyik-Trawogerc, 2014; Beltrão de Lucena Córdula, 2015). Denzin (2009) catalogs this reality as a hostile research environment. For some researchers, it is uncritical how biomedical ethics has been translated into guidelines, protocols, and criteria for evaluating ethics in health research or human behavioral studies (Gergen, 2013; Tuck and Guishard, 2013; Guerriero and Castaño-Pineda, 2015). In some cases, the constitution of institutional ethics committees facilitates the evaluation of ethics in the research project being considered for approval, and in others, it does not (Parsella, Amblerb, and Jacenyik-Trawogerc, 2014). In educational research, the positioning among researchers and educators is that ethics in educational research must be understood in the cultural and institutional context of individuals and their communities (Davis and Holcombe, 2010). Ethics in the social and educational world has many faces and nuances. This is essential for educational researchers worldwide who work with communities and marginalized people, and from cultures of poverty, to bring justice to the world (Guerriero and Castaño-Pineda, 2015). This concept of ethics differs from the vision of biomedical ethics currently applied to the evaluation of studies submitted for approval by institutional ethics committees. This possibility can be realized through protocols that are more sensitive to social and educational research (Denzin, 2009). This possibility can be facilitated by the constitution of institutional ethics committees with professionals knowledgeable or akin to social and educational research (Parsella, Amblerb, & Jacenyik-Trawogerc, 2014). Educational researchers need to bring a more consequential discussion on ethics in academic research to professional forums if we aspire to a broader scope and achieve social justice through social and educational research (Guerriero and Castaño-Pineda, 2015).

10

The Use of Standardized Tests to Measure Learning: Past, Present, and Future

Ángel Canales Encarnación

Measuring learning through tests has been a common practice in the field of public education since its inception. The use of standardized tests to measure knowledge has been a methodology that has evolved alongside educational research in explaining the educational quality of academic programs, schools, school districts, and countries. Since the mid-20th century, the quality of education worldwide has been measured based on students' scores on standardized tests. Explaining a student's scores on a standardized test has involved a profound reflection on the phenomenon of education and the reality and context in which it occurs. The effectiveness of standardized tests to measure and improve educational quality has been the subject of criticism and debate (Ponce, 2014b; Ponce, Gómez Galán, & Pagán-Maldonado, 2018). This chapter examines the use of standardized tests as an alternative methodology to educational research methods. The chapter is organized into three sections: Standardized tests past, present, and future.

Standardized Tests for Measuring Learning: Past

Where can we begin to learn about a widely popular concept in the context of educational evaluation in today's society? Somehow it is necessary, without trying to overwhelm the reader, to resort to some background information and data that allow us to identify the starting point for understanding the concept and its impact on the evaluation processes that prevail in education today, especially when it comes to evaluating the quality of school systems. In this account, you will recognize the incorporation of previous learning that constituted part of your training as a future teacher, mainly in the Teacher Preparation Programs of your corresponding University. The concern for knowing, measuring attributes, and predicting the behavior of human beings is an issue not only associated with the work of science but also with education and the social sciences since ancient times. In education and the social sciences, unlike the natural sciences, we cannot directly measure the attributes of interest associated with student learning. We resort to measuring these attributes indirectly, usually by elaborating instruments that we eventually recognize as tests. We will make inferences or interpretations of their results to determine whether or not the student learned the material of interest that is the object of evaluation. This type of test corresponds mainly to those that each teacher performs internally in his or her classroom to determine the day-to-day learning associated with the teaching and learning processes. However, in the evaluative practice, the teacher handles internal classroom tests. Still, his approach is impacted by external tests (usually constructed by external professional companies), which we call standardized. These tests have different uses, but we see their results mainly to corroborate student learning, monitor the quality of teaching in schools, determine the quality of school systems, and, in particular cases, their incorporation to determine the teacher's professional performance. I anticipate that the uses and interpretations that we are accustomed to hearing and that have been legitimized in the current evaluative educational discourse do not necessarily imply the correct use and performance of this type of test. This is something that we will be able to clarify and illustrate throughout the content of this chapter.

Although the use of standardized tests began in the 1990s (Linn, 2000), it is recognized that the introduction of this type of test at that time was something of a novelty but, at the same time, a severe challenge to the organizational culture especially in universities. The concern for determining the learning obtained by students through a process known as **measurement** can be seen reflected throughout history in various initiatives. The use of the oral formulation of the question to determine the knowledge obtained by the student stands out in terms of its relevance to the current educational context. E. L. Thorndike is credited with having introduced the objective question in written achievement tests (Cirino, 1984). It is possible that from this event in history, the teacher's practice related to the writing or formulation of questions in exams to corroborate the learning of students in certain subjects, the product of teaching, could begin.

Throughout history, there is evidence of development mainly of a theoretical and statistical nature related to the original concept called testing. This opened the way to what, in education, we call today psychometrics, or the science that deals with making measurements on the multiple aspects of human behavior. According to Aragón (2015), the origins of this science seem to date back to China, where there is evidence by the emperors of evaluation processes to examine officials regarding specific jobs to be performed and the existence of evidence associated with the concern to evaluate officials in particular competencies. Reviewing previous learning from your training as an educator, mainly those of a philosophical nature, you will remember the figure of the philosopher Plato. In his text *The Republic*, he recognizes that no two people are alike. This philosopher acknowledges that human beings vary in their qualities and attitudes.

For this reason, *The Republic* was made up of three types of people: the philosophers (governors/wise men), the people (artisans and farmers), and, finally, the warriors (represented by the firm and brave). For his part, Aristotle, Plato's disciple, recognized mental and moral differences in individuals, mainly according to the social level, race, and sex. Therefore, it is evident that the concern for measuring the attributes of the human being and the learner had its beginnings at the dawn of history.

It is recognized that at some point in history, the field of psychometrics stopped calling this basic unit of the test a question and instead called it an item (Cirino, 1984). This was mainly due to the great boom of measurement instruments, which did not necessarily continue to respond to the original conception as a sentence with its corresponding signs. The introduction of new forms of scale, interest inventories, and adopting the item's notion in the instruments used is recognized as part of this new variety. Some other antecedents identified in history associated with the development of tests or examinations, mainly corresponding to ancient times, identified in history through the analysis of some books, confirm the existence of Chinese examination systems as well as the practice of the Greek Empire of conducting examinations to determine the mastery of physical and intellectual skills; likewise, the incorporation of the Socratic method based on the formulation of the question, which was adopted in the Middle Ages for the entrance to the Universities.

By the year 1890 (19th century), the word *test* appears for the first time. At the beginning of the 20th century, a flourishing of this science (psychometrics) is observed with the development of tests that currently correspond to measuring various psychological attributes related to mental capacity, achievement, personality, and vocational interests. Hogan (as cited in Aragon, 2015) recognizes Alfred Binet of France as the Father of Intelligence Testing. The development of this first intelligence test also introduces the first measure of a standardized nature in psychometrics. With it, it is possible to identify children in schools, initially in Paris, in need of Special Education services, as a product of a lower IQ than other students measured with such instruments.

In the modern era corresponding to the 20th century, some approaches and developments to testing can be recognized in the figure of Binet and Simon. They developed the first intelligence test (Aragón, 2015). Likewise, during the First and Second World War, the development and use of collective tests of intelligence and emotional aspects are evidenced.

At the beginning of the 20th century, the development of a considerable amount of theoretical and statistical writings and publications related to standardized achievement tests is palpable. E. L. Thorndike stands out, mainly his concern for the scientific nature of education and his efforts to increase the measurement to achieve reliability of the results by the corrector or scorer of the measuring instruments.

History recognizes the *Stanford Achievement Test* in 1923 as the first standardized achievement test. Finally, in our attempts to connect some previous learning from our training as teachers with the origins and understanding of standardized tests, it is also significant to attribute to the figure of E. L. Thorndike, his contribution and contributions to Learning Theories, mainly connected to the development of this type of tests and learning theories of behaviorist nature. These theories recognize that learning is primarily composed of stimuli that provoke responses, which can be measured objectively through measuring instruments such as tests.

This conceptualization allows us to recognize that any particular question or item formulation of a *test* or examination is seen from behavioral learning theory. This specific stimulus elicits a learning response from the student. This response is operationalized in the written *test* or quiz, whose correction constitutes the evidence to interpret the learning obtained and eventually, mainly through global student results, to make inferences associated with the quality of students, schools, and school systems. Students give the answers, the aggregated answers of groups of students, and their eventual use to infer the quality of schools and educational systems are some of the main controversies underlying current evaluative practice.

Standardized Tests for Measuring Learning: Present

The use of tests to determine students' academic achievement in the classroom is not a strange matter for the educator. Their practice is greatly influenced by aspects related to evaluative processes. However, nowadays, other types of measurement instruments have been incorporated; particularly, **standardized achievement** tests whose use and interpretation of results have repercussions on the quality of student achievement and the quality of schools, school systems, and public opinion in general. Hence, the importance for the future educator, as the one active in teaching, is to engage in a deep critical academic reflection on this type of test or measuring instrument. Renowned scholars associated with the field of education and particularly with the measurement of student academic achievement at the local and international levels have insisted in their writings on the importance of educating and promoting greater dissemination on the correct use and interpretation of the results associated with this type of test (Popham, 1999; Cirino, 1984). There is no doubt that the use of written examinations, particularly those of a standardized nature, has a significant impact on education, which extends beyond the particular results obtained by students. The implications and repercussions of the products and interpretations of these tests include determinations on the quality of schools and even on the quality of the teacher who teaches, not to mention their aggravating biased contribution to shaping public perception of the quality of education. In fairness, it is recognized that there is room to encourage the correct use and interpretation by teachers and school authorities and their possible contribution to improving the quality of education.

Student achievement and the quality of education in schools is always a hot and passionate topic. However, in recent decades, we have observed the influence of legislation that favors the enhancement of this type of tests to measure students' academic achievement. Past laws such as *No Child Left Behind* (NCLB) and others, whose application was extended to Puerto Rico as a territory of the nation and other states, establish measurement mechanisms through standardized tests to determine the quality of education and to account for the results of students' academic achievement, including the establishment of sanctions, classifications, and determinations in teacher performance. Some of the documents or laws that are recognized to have driven this practice are the *National Assessment of Educational Progress* (NAEP) to measure student achievement and illustrate trends in student performance over time (1960) and the *Elementary and Secondary Education Act* (ESA), passed in 1965 by the U.S. Congress, which required the evaluation of educational programs that received funds under Title I (i.e., programs aimed at students from low-income families) (Medina, 2007). Some of these tests, such as the NAEP, continue to be used today in the nation's territories such as Puerto Rico and other states.

Still fresh in the minds of educators in Puerto Rico and other states in the nation is the image of the implementation in the past decade of the *No Child Left Behind* legislation, whose implications contributed to the classification and estimation of many schools in the country. According to Koretz and Hamilton (as cited in Medina, 2007), the NAEP and ESA requirements constitute the first uses of standardized achievement tests as a tool for monitoring the performance of the American student body. Currently, we have the Every Student Succeeds Act (ËSSA) signed in 2015, which also continues the practice of using standardized tests as accountability mechanisms, even in some localities or territories such as Puerto Rico, extends its system of implications or sanctions to the quality of teachers' professional practice, as it is considered part of their professional performance. There is no doubt that the use of this particular type of test responds to the demands of sectors of an economic and political nature, hence their legitimized use in the context of today's society.

What Are Standardized Tests and the Scope of Interpretation of Their Results?

What better way to begin the approach to such a controversial concept in education than by approaching a definition of standardized tests. The term "standardized" refers to applying, scoring, and interpreting a standardized test (Cirino1984; Medina, 2007). The test is accompanied by a manual that includes information and technical considerations related to its development, specific instructions for its application, and, in particular, statistical relations for the interpretation of its results. It should be pointed out that although, at present, some degree of teacher participation is usually sought in the process of developing these tests, it is professionals specialized mainly in the field of psychometrics who develop these tests, taking into consideration particularities of the curricula in their respective localities, populations and their particular characteristics, and other considerations that mainly contribute to their validity (seeking to measure the educational content of interest) and reliability (seeking the consistency of the results once they have been applied). It is essential to recognize the technical rigor demanded by all types of standardized tests and establish norm or standardization processes, which will be used to interpret the results obtained by students. Regularly, their use is based on making comparative interpretations with said norms or reference groups and not, as we are used to seeing nowadays, mainly as an indicator of the educational quality of schools and school systems.

When interpreting standardized tests, established norms or criteria are used; so this type of test is often referred to as a standardized test. A standardized test exposed to a normalization process is one for which test norms have been calculated based on the responses of a representative sample of people (Medina, 2007). Standardized tests are usually administered to large samples of students with different characteristics, including gender, age, school level, socioeconomic level, race, residential area, and other variables of interest. Once administered, the distribution and behavior of the scores obtained are studied, leading to the determination of norms and statistics that correspond to the typical performance of the group examined. The interpretation is a relative one taking into consideration the standards established with the reference group. Obtaining the result or score of a given student is meaningless if it is not interpreted in the light of the comparison with the reference group. Before any numbers of questions are answered correctly by the student, we must refer to those answered correctly by the rest of the group as a first step to approach a correct interpretation of the results. The importance of understanding is relative and not absolute. Usually, the score obtained by the student is converted or transformed to some convenient scale to facilitate its comparison. In education, the teacher is familiar with the use of a scale called the percentile rank. To illustrate the use and interpretation using this scale, we can consider a student with a score of 20 correspondings to a percentile rank of 90. The understanding leads us to believe that the student obtaining this score exceeds 90 percent of the students in the reference group. These scores obtained by the reference group establish the patterns or norms of comparison that allow us to infer whether the scores were of a high, average, or low nature by using statistics corresponding to the level of measurement of the variables.

In current practice, there are often conflicting positions among educators regarding the correct use in interpreting and using the results of standardized achievement tests to make decisions regarding the quality of teaching (Cirino1984, Popham 1999, Medina 2007) and particularly school systems. It is common to observe in standardized aptitude tests the use of these results to predict students' future behavior in terms of their academic performance at the next level of education. Popham (1999) insists on using standardized tests to make comparisons, as we pointed out in the previous example, but not to make evaluative judgments regarding how good or bad a school is. The leading critics of this practice in the interpretation of school or educational system effectiveness results base their conceptualization on the fact that the achievements considered represent a sample of the learning obtained by the student within a universe of possible knowledge. Of course, there is also the concern that the content selected for such tests should be equally representative of the teaching processes imparted in each school and particularly to the students.

At this stage of the conceptual understanding of standardized tests, it is convenient to review the theoretical considerations on the concept of test in education, which probably underlies most of the main controversies that have been identified regarding the use and interpretation of standardized achievement tests. The test in education, particularly from measurement processes, is a representative sample of items related to the content to be measured (Medina, 2007). For this reason, they measure a limited portion of the knowledge imparted to the student in the teaching process. It is necessary to delve deeper into some of the main factors attributed to the incorrect use of these results to measure educational quality.

What Are Some of the Main Controversies for Declining Standardized Tests as an Indicator of Educational Quality?

Representativeness of the Content

Although psychometricians in charge of test design perform rigorous procedures to ensure the representativeness of the study content within the test universe, there is always content of study that may not be incorporated into the test. Scholars who support this concern argue that items can always be incorporated into the test that is not aligned to the curricular content of the study and particularly to the teaching processes.

Selection of Items in the Test

As we have emphasized, the test is a representative sample of items related to a universe of content. The final selection of items that qualify for the test assembly are those that we call in psychometrics of medium difficulty, that is, items that, by their nature, can discriminate, being only answered by 40−60 percent of the students, thus allowing discrimination among the students tested. Many contents that may be experienced in classrooms and for which items have been contemplated in the test question bank, which have been answered correctly by more than 90 percent of the students, are eliminated from the test because they cannot discriminate among them, examinees.

Types of Learning That Are Examined

From the curricular point of view, we can ask ourselves what type of learning is valuable to incorporate in a standardized test and whether such learning guarantees equality for all students in terms of the opportunity of exposure to content at school to ensure their learning about it. According to Popham (1999), standardized tests incorporate as part of their items contents that represent learning that the student acquires outside school. Logically, considering the reality that the student's socioeconomic level influences the educational context, many students in low socioeconomic schools may be deprived of rich learning experiences that are not incorporated into regular teaching practice. Today, we might call this factor exclusion from meaningful learning in teaching. In the same way, items that, due to the intellectual nature of a particular student, he/she is limited in answering are incorporated in this selection of learning. There is a diversity of intelligence, as established in recent learning theories. Still, not all of them are included or there are items representative of them in the construction of standardized tests. Likewise, the contents of subjects that often do not respond to the economic demands of society, mainly those associated with the humanities, are excluded, giving prominence to the contents of issues such as mathematics and science.

Returning to some of the considerations outlined in the introduction, the proliferation of standardized achievement tests is evident with the evolution of society and technological advances. Nowadays, the intensification of psychometric companies worldwide shows that the issue of standardized tests transcends the classroom and has a strong impact of an economic and political nature, based on the so-called "large-scale" evaluation systems, which encourage their use and practice to legitimize the quality of school systems and establish sanctions, to produce educational reform processes. The current help and experience with such tests as an accountability mechanism from which we must learn leads us to ask: Have these mass standardized testing initiatives and the accountability processes based on them produced any significant change in education? Have the particular scores obtained by schools and the classifications or *rankings* obtained and disseminated to the detriment of educational quality delivered any change in their educational practices? Martínez and Prada (2012) point out that the North American experience shows that linking standardized test results to penalties for teachers and schools is an insufficient strategy to motivate positive changes in the educational system. There is much to be researched and many factors to consider when determining student learning and the quality of educational systems. Perhaps, the behaviorist vision of education has focused on scores as the answer and the standardized test as the primary stimulus. Still, the learning process is much more complex and multifactorial. Perhaps, without totally delegating quantification in the measurement processes, we should initiate and consider space for research processes of a qualitative and mixed nature to consider making contributions that are truly indicative of learning and, therefore, of our schools' educational quality students.

Standardized Tests for Measuring Learning: Future

What to expect as an educator in the future regarding the use and interpretation of standardized tests to measure the quality of school systems and the academic achievement of students? How to obtain the correct direction in terms of the understanding and proper use of these results to bring about a fundamental change in the path of improvement in the quality of education and the specific learning obtained by students? What particular role should the various actors responsible for education assume in terms of the correct use and interpretation of these results? Undoubtedly, the education that we seek in this article through the proliferation of indicators that allow the understanding and proper application of the results in educational contexts is a starting point, in my opinion, of outstanding contribution. It is necessary to illustrate and emphasize the correct discourse in the use and interpretation of these results. As reflected in this article, the existence of standardized tests and their use goes back to the origins of history. Its roots and application may respond to diverse considerations. Still, the political pressure based on current laws that support the use and interpretation of standardized tests to recognize the quality of school systems stands out. The use of these tests as an indicator or predictor of future learning, which allows for sustained monitoring of each student's performance and academic potential, stands out. However, the main critical issue highlighted in this paper and the central reviews consulted as part of the literature review is the frequent and widespread use of standardized tests as a barometer or indicator of the quality of both school systems and student academic achievement.

The reasons that constitute the cons and examples of why such practice should be discontinued have been briefly presented in the content of this paper. It is convenient as part of the future conceptualization to alert the educator to look at the alternatives that their professional practice can contribute to this issue. Providing meaningful learning experiences for students in school contexts can be a starting point that contributes to a better understanding of the quality of education and student academic achievement. Through objective evaluation processes, we can obtain indicators that allow us to better estimate learning outcomes. Verdejo and Medina (2008) point out some of these attributes of assessment as a complex, systematic, and comprehensive process that allows making judgments about student achievement by emphasizing that "through assessment, teachers have the power to collect qualitative and quantitative information that, among other things, can: change the curriculum, determine teaching methodology, shape the processes of learning and teaching, influence the self-concept of students, shape their academic future, influence their self-esteem and control their access to different academic opportunities" (p. 44).

Considering the various positions highlighted in the article, which show points that favor or disapprove the use and interpretation of standardized tests to measure the quality of education, we wonder about the route to follow to identify correct indicators or at least approximations that allow us to produce significant changes in the teaching and learning processes experienced in schools. In this regard, Linn (2000) points out not to put all the weight of the grade on a single exam but to use multiple indicators that increase the validity of inferences about student achievement. Most educators agree on the importance of the test incorporating meaningful learning that constitutes students' actual learning achievements. In the search for accurate indicators of educational attainment and quality, the significant contributions of the pedagogical field should not be left aside, mainly from the contributions of constructivist and sociocultural theories, which report great advances in terms of the understanding of learning, human intelligence, educational practice, school and professional cultures, and development and needs of students, which have allowed the promotion of curricular proposals and broader, more open, and flexible evaluation models (Campos, 2011; Carretero, 2011; Coll, 2010) that contribute to the development of higher-order intellectual skills, as well as the acquisition and promotion of desirable values and attitudes, which make the individual an autonomous, critical, reflective, participatory, supportive, and responsible subject for his peers (Moreno, 2002, pp. 32-33).

Recognizing that the educational activity is one of a complex and multifactorial nature should make us look at results that are more directed to learning processes than products obtained from immediate test results. Without discarding these as part of these multiple indicators, the educator must direct the teaching and learning processes to those that propitiate a reflection of higher quality and offer a panorama of education of more excellent acquisition of competencies by the student, considering a level of critical and higher-order thinking in each student. This demands each educator an invitation to transform each classroom into learning environments rich in learning experiences, which appease the opportunity that each student has to demonstrate what he/she knows and is capable of doing as a result of his/her learning.

There is a need to deepen the understanding of student learning through qualitative and quantitative research approaches (using triangulation methodologies). Considering the exclusive use of standardized achievement test results and any other measurement instrument to determine student academic achievement and make critical decisions is a grave ethical and technical mistake (Medina, 2007). Herman and Golan (as cited in Hernández Madrigal, Ramírez Flores, & Gamboa Cerda, 2018) point out that if the test results represent meaningful learning, then the tests can contribute to the achievement of the objectives. Still, if the results do not mean education, such practice can be considered a misguided tool.

As educators, we have to contribute to the correct dissemination of the use and interpretation of the results of standardized tests. Perhaps, the paradigm shift away from seeing these tests as the sole and successful potential of the quality of education and academic achievement of students constitutes a significant contribution of educators, mainly by educating the parents of each student and other members of the school community about the correct use and interpretation of these results. Waking up from the legitimization that test companies and local and international governments and laws have established by relying on test results is a major challenge. But as I indicated in the introduction of this article, I bet on each teacher as an educator to transform this vision and redirect the gaze to other learning indicators that contribute to the understanding of each student's learning.

As educators, we must ensure that the inferences made about each student's learning are based on valid, reliable instruments and, above all, on the recognition of learning as multifactorial, incorporating multiple indicators and evidence of the processes, which lead to a reflection of the learning that promotes the quality of education and the academic achievement of each student. Redirecting the focus and establishing proposals for evidence through comprehensive *assessment* systems could make a difference in the quality of education. After all, we must ask ourselves if any significant results have been achieved through *accountability* processes based exclusively on legislation and penalties in light of standardized test results alone. Why not walk the walk? Why not give the procedures and evidence that occur in each classroom a greater educational recognition? Inviting teachers to reflect on this process and identify appropriate evaluative tools significantly recognizes their work and an excellent mechanism for their professional development. At the end of the day, if we do not manage to impact changes in teaching and learning processes and the quality of what we identify as evaluative practices, it will not be easy to improve the quality of education. I bet on the teacher, in the range of proposals that can be conceptualized before the debate and controversies, that this article managed to transmit to every educator, administrator, parent, and school system working to achieve the concept of educated man that today's society demands.

11

Research-Evaluation of Learning

Omar A. Ponce, José Gómez Galán, and Nellie Pagán-Maldonado

Learning is considered the main responsibility of educational systems (Ponce, 2014b). Research-evaluation of learning has been an essential topic in the field of educational research since its inception. Research-evaluation is defined as the use of research methods to evaluate educational programs, policies, procedures, and institutions (Ponce, Pagán-Maldonado, & Claudio-Campos, 2017). At the beginning of the 21st century, the field of evaluation research is much more sophisticated and complex than in its early 20th century beginnings:

1. **Accountability**. At the beginning of the 21st century, accountability for learning is a legal requirement that many governments worldwide demand from their educational systems in the interest of improving society and its quality of life (Escudero, 2016). The concept of accountability is used to describe the exercise that each educational institution must undertake to generate data on student learning and use these data to improve the quality of education (Ponce, 2014b).
2. **A more precise understanding of the complexity of learning**. With accountability emerges a clearer understanding of the complexity of education and the challenges of promoting and determining learning as an institutional product (Ponce, 2014b).
3. **The sophistication of institutional learning assessment research.** At the beginning of the 21st century, three models dominate institutional learning assessment research: standardized testing programs, institutional learning assessment, and institutional assessment research programs. These models are different in their rationale and approaches, but they pursue understanding and improving student learning.

In this reality, research-evaluation emerges again as an essential instrument for accountability and determines the effectiveness of programs and educational institutions in promoting student learning (Escudero,

2016). At the beginning of the 21st century, more and more educational researchers are bringing their methodologies and research strategies to evaluating educational programs and institutions (Escudero, 2016). This chapter discusses the state of the art of educational research-evaluation in accountability processes to determine the effectiveness and quality of educational institutions.

Accountability

Accountability is determining and evidencing with data the effectiveness of schools and colleges in promoting student learning and academic quality. This information can be used to inform all audiences interested in the educational institution, such as accrediting or licensing agencies, administrators, students, parents, donors, or companies that recruit graduates. The accountability exercise should provide data that can be used to determine the effectiveness of the school's or college's academic programs, student services, or educational practices. An essential political exercise of accountability is to demonstrate students' academic success concerning their fiscal cost and the public money they receive for their operation in the form of financial aid and student scholarships (Ponce, 2014b).

The accountability movement began to take shape in the 1980s, with neoliberal philosophies entering public administration in many countries worldwide. In the United States of America, these philosophies came to public education in the face of growing concern with academic quality, the high costs of education, and the importance of a well-educated workforce for insertion into a global, technological economy. Accountability mechanisms in the field of education are implemented through the following administrative structure (Ponce, 2014b):

1. **Quality standards.** It consists of identifying the expected learning by the subject of study, and that students should master in the different academic levels of their education.
2. **Accreditation standards**. Quality criteria that every K-16 educational institution must meet to be recognized as accredited.
3. **Evidence-based professional practice.** The effort to promote a scientific culture in decision-making based on research data and scientific evaluations of learning.
4. **Scientific methods for producing data and evidence**. The use of systematic and rigorous research and evaluation methods to create data is used as evidence to guide all educational practices.

The accountability movement brought to the discussion the complex and sometimes difficult to differentiate interrelationship between measurement, assessment, evaluation, and learning research when talking about decision-making and the evidence to be used for these purposes. Although many people in the field of education use these concepts as synonyms, others do not because they claim that there are theoretical and practical differences between these concepts. For example, the measurement of learning is associated with the construction of test items and the study of the student's response to them to make inferences and decisions related to the student's diagnosis or performance, placement in an educational system, promotion, or evaluation. Assessment is defined as collecting, analyzing, and interpreting data about learning, a program, or an institution for improvement. Assessment means determining the worth of something. This can be an educational policy, a service, or an academic program. This discussion is the recognition that there is more than one mechanism for collecting data on learning and that all contribute to decision-making in improving education. There also emerges a recognition that the disciplines of measurement, assessment, and evaluation have been distant from each other. There is a need to build bridges that connect this diversity of mechanisms for collecting learning data to improve the accountability process (Secolsky and Denison, 2014).

Accountability for learning brings to the fore the complexity of educational institutions, be they primary, secondary, or university, and the challenges faced by educational researcher-evaluators. For example, learning in an educational institution can be approached from different perspectives, such as knowing what occurs in a classroom, learning that occurs in an academic program, learning that occurs in student support services (mentoring, advising, and tutoring), and learning an institutional product. For accountability purposes, the call from the political world to administrators and educational researcher-evaluators has been to determine learning as an institutional product (Ponce, 2014b; Ponce, Pagán-Maldonado, Gómez Galán, 2017). Accountability for learning as an institutional product raised the following questions for educational researchers: (a) What data are needed to understand learning? (b) What instruments should be used to collect data on learning? (c) Who collects the data? (d) What procedures are used to collect the data? (e) How are the data interpreted? (f) Who uses the data? (g) How are the data used? (h) How are the data reported (Ponce, 2014b)? This, in turn, has propelled interest in institutional research (Escudero, 2016).

Learning as an Institutional Product

There are three premises about "learning as an institutional product," which help educational researcher-evaluators understand the logic and procedures for institutional improvement and accountability purposes (Ponce, 2014b).

Learning as an institutional responsibility. The starting point of accountability is the institution's learning in its mission and its goals and objectives. The mission of each institution declares its particularities, which, in turn, differentiates it from others. The institution is accountable for the structure it adopts to educate its students.

Learning is a shared responsibility of teachers and students. Another point of reference for accountability is the curriculum. The institution must provide evidence that the graduating student has the competencies stated in the mission statement (learning outcome assessment). Institutions are urged to examine the alignment of their curricula to connect courses to the desired competencies. In this view, student learning in an institution is predominantly behavioral and linear. Students are planned and then exposed to educational experiences and activities to elicit the desired learning (institutional product). Learning as an institutional product implies that both teachers and students are responsible for this process and dynamics.

Learning as an institutional intention. The third benchmark for accountability is to use the data to improve the learning structure of the institution: its curricula, programs, and services (e.g., Allen, 2004; Asting and Linsing Antonio, 2012; Middaugh, 2010; Kramer and Swing, 2010; Suskie, 2009). Therefore, if consistently and systematically assessed, it is assumed that the institutional structure of learning can be improved. This increases the chances of positively impacting the institution's effectiveness in promoting student learning. From this perspective, learning is an institutional product that is provoked.

The accountability movement brings to the discussion that institutional learning has social and psychological aspects. The social processes of learning are the activities adopted by the institution to socialize or relate the student to the content, values, and skills of the profession to which he or she aspires in the case of a university or the learning that is aspired to develop in the case of an elementary school. Explaining learning as a product and an institutional responsibility implies analyzing the structure used by the educational institution to induct the student into academic life, provoke students' learning, and develop them until they complete their studies. The aspects that define the **learning structure of** an educational institution are the following (Ponce, 2014b).

Curriculum. The curriculum is the study plan to provoke learning. The curriculum identifies the learning to be developed in the student (curricular content), outlines the sequence to be used (order of courses), and the teaching and assessment techniques employed to determine achievement.

Academic and institutional policies. Curricula are implemented in the context of the institution. Policies are the regulations adopted by the institution to regulate the operation of its educational programs. Institutional policies shape the curricula or institutions in aspects such as admission criteria, student academic progress or development, learning assessment criteria, student and faculty responsibilities and duties, and criteria for financing studies.

The institution's teaching philosophy. Since the 1990s, more universities have become aware of student diversity in their classrooms. This diversity began to be channeled through modes of study (e.g., traditional 18−22-year-old college students, working adult study programs, professional in-service study programs, and online programs) that responded to that range of lifestyles and realities. Student diversity highlighted subtle but significant differences in the relationship that was to occur between a teacher and a student. For example, the role of an adult educator requires a different approach to teaching than that of an 18-year-old student. The response of many universities was to define the teaching philosophy that the institution wanted to occur in the relationship between students and teachers in the classroom and outside. For example, some universities state that the role of the faculty is one of a facilitator and that the relationship between teachers and students should be one of professionalism at all times. As administrative tools for regulating practice and the relationship between faculty and students, teaching philosophies transcend the traditional view of the contractual duties and responsibilities of faculty. Teaching philosophies become guiding principles and codes of behavior expected of faculty in their relationships with students.

Student support structures. It means the scope and organization of student support services to address and respond to the academic and personal needs of the student: tutoring, psychological counseling, or spiritual counseling.

Learning as an institutional product is also examined through the lens of the learner and his or her psychology. The psychological aspects of learning involve understanding how the learner perceives and experiences the social processes of learning. For example, whether or not they like them or perceive them as relevant. The premise is that how people interpret their life realities determines their behaviors. If you understand these perceptions of learning, you can understand their behavior and intervene with them. Common research-evaluation strategies in this regard are to understand the following psychological aspects of learning: (a) students' motivations for their studies, (b) students' perceived relevance and pertinence of their courses, (c) levels of satisfaction with their programs of study or the institution in general, and (d) processes of adaptation to university life.

Research-Assessment Strategies to Determine Learning as an Institutional Product

Measuring, describing, explaining, documenting, and improving learning as an institutional product are at the core of the accountability movement. Three models dominate and translate notions of learning into concrete actions and institutional data collection programs for accountability (Ponce, 2014b): standardized tests, learning assessment programs, and institutional research programs.

Standardized Testing Program

The focus of this strategy is the construction and administration of standardized tests to measure learning. The results are analyzed, and mastery levels are determined according to the student's correct answers. It is a psychometric model because learning is defined based on the acceptance of mastery criteria on which the test is based. In the United States of America, this learning assessment model emerges at all general educational levels as a requirement of the Federal Government to measure and compare the learning between academic institutions and allocate funds in light of the results. The use of standardized tests in the field of education can be traced back to the work and contributions of psychology at the beginning of the 20th century and the measurement of intelligence attributes on students to better serve them academically. At the beginning of the 21st century, standardized tests to measure and compare the learning of institutions and countries are visible around the world. As a model of learning research-evaluation and accountability strategy, its implementation follows the following logic.

Phase I. Identification of learning. Learning is determined by quality standards or by groups external to the educational institution. Quality standards are imposed through institutional accreditation agencies or professional requirements that define the expectation to be achieved.

Phase II. Data collection. It focuses on the administration of the standardized tests as data collection instruments. Coordination of the test administration process falls to the institutional assessment and institutional research offices or the contracting of external companies that build the test and administer it.

Phase III. Data analysis. The tests are analyzed by the institution's statistical or appraisal offices or by external companies that design and administer the tests.

Phase IV. Evaluation of the improvement. The accrediting agency or external evaluator sends the results to the institution. It is up to the faculty and administration to discuss the results, identify intervention strategies, and implement them to respond to the accrediting agency and improve.

Learning Assessment Programs

There is no universally accepted definition of learning assessment. This means that there are no defined and accepted positions on its practice (Kramer and Swing, 2010; Secolsky and Deninson, 2012; Ponce, 2014b). When writing this book, it is easier to talk about assessment and describe its practices and methodologies than to define it (Ponce, 2014b). Learning assessment is considered an educational movement, a discipline of study, and a developing field of employment. Two positions dominate the literature when explaining what learning assessment is. One position defines it as a systematic process of collecting data on student learning (Huba and Freed, 2000; Kramer and Swing, 2010):

"Assessment is a systematic process of gathering, interpreting, and acting upon data related to students learning to develop an understanding of what students know, understand, and can do with their knowledge as a result of their educational experience. The process culminates when assessment results are used to improve subsequent learning" (Huba and Freed, 2000. p. 8).

A second position defines it as research applied to the study of learning (Astin and Linsing, 2012; LeCompte and Aguilera-Black Bear, 2012):

"Good assessment is excellent research, and the ultimate aim of such research should be to help us to make better choices and better decisions in running our educational programs an institution" (Astin and Lising Antonio (2012).

For Suskie (2009, p. 13), the appraisal is more similar to action research because of the cycle of phases used in its implementation, where data on learning are collected, and decisions are made in light of this information:

"....assessment differs from traditional research in its purpose and therefore in its nature. Traditional empirical research is conducted to test theories. At the same time, assessment is a form of action research, a distinct type of research whose purpose is to inform and improve one's practice rather than make broad generalizations...The four-step assessment cycle of establishing learning goals, providing learning opportunities, assessing student learning, and using results mirror the four steps of action research: plan, act, observe and reflect. Like any other form of action research, assessment is disciplined and systematic and uses many of the methodologies of traditional research."

The two positions agree that assessment is about collecting data on student learning to understand and improve it.

According to Earl (2003), the assessment does not seek data to pass judgment on student learning, as is done in learning assessment. Data are collected to describe what the student can and cannot do with learning and use this to work with their deficiencies and strengths. Allen (2004) describes the role of assessment in an educational institution as follows:

"assessment is a framework for focusing faculty attention on student learning and for provoking meaningful discussion of program objectives, curricular organization, pedagogy, and student development" (Allen, 2004, p. 4).

In this chapter, the term assessment refers to collecting, analyzing, and interpreting data on student learning. This seems to be the dominant position in the literature:

"Assessment, at its heart, is about the collection, analysis, and interpretation of information related to a particular issue or outcome of interest" (Secolsky and Deninson, p. xviii, 2012).

There is a need for two types of assessment to demonstrate student learning: outcome assessment (outcome assessment) and assessment of educational processes (formative assessment). According to LeCompte and

Aguilera-Black Bear (2012), the evaluation of products answers the "what" of learning but not the "why." This assesses the educational processes necessary.

Assessment of the outcome or product. A learning product is that students should demonstrate behavior, knowledge, or skill as a result of their academic experiences. The primary emphasis of accountability is to show that students upon graduation have mastered the knowledge and abilities stated by the institution in its mission, goals, and objectives. The product assessment focuses on collecting data to demonstrate or evidence student learning as an institutional outcome.

Assessment of the educational process. It means collecting data to understand the institution's educational methods and how learning occurs (Allen, 2004). According to Forsythe and Keith (2004), the assessment of the educational process is essential because it brings to the discussion the institutional context where learning occurs: the institutional processes, its culture, dynamics, relationships among faculties, and facilities. In this way, data are collected to understand the context of learning and how it affects the product.

To achieve a learning assessment that facilitates institutional improvement and accountability, it is necessary to identify the data needed and plan how it will be collected. Astin and Linsing (2012) put it this way:

"Because assessment and evaluation should thus be designed to improve decision making, and because decision making inevitably involves causal reasoning ('Alternative A is chosen over Alternative B or C because we believe that it will lead to a better outcome'), assessment results are of most value when they shed light on causal connections between educational practices and educational outcomes" (Astin and Linsing Antonio, 2012, p. X).

A dominant position on the appraisal is that there are many ways to implement (Kramer and Swing, 2010; Secolsky and Denison, 2012):

"We acknowledge that there are no one-size-fits-all solutions for successful learning outcomes assessments. Just as learning is complex and often messy, so are efforts to measure and evaluate it" (Kramer & Swing, 2010, p. 4).

Assessment is conceived as a systematic process of collecting data on student learning. Its methodology is captured in a four-phase cycle, explained below (Suskie, 2009; Allen, 2004; Huba and Freed, 2000; Kramer and Swing, 2010). The names of these phases vary among authors, depending on whether the assessment applies to the classroom, an academic program, or an educational institution. What seems consistent among various authors is the objective that each phase pursues as part of the assessment cycle (e.g., Suskie, 2009; Allen, 2004; Huba and Freed, 2000; Kramer and Swing, 2010).

Phase I. Identification of the learning to be assessed. The assessment cycle begins when the institution defines the learning objectives for its programs and courses. The objectives should clearly state, in measurable or observable terms, the learning that students should demonstrate. In this way, different teachers assessing students' work should come to a similar conclusion as to whether or not the students achieve the expected objectives.

Phase II. Data collection. The second phase of the cycle consists of developing and implementing the learning assessment. Instruments and data collection procedures are developed and validated. Implementation is articulated, and data are collected. This phase of the assessment cycle was explained in Chapter V.

Phase III. Data analysis and decision-making for improvement. This phase consists of interpreting the assessment data and converting it into actions to improve learning and the institution and its academic programs. The most common strategy for analyzing the data is to use computer programs to handle the volume of information collected from students. The results are discussed and interpreted in faculty and administrative meetings. The purpose of these meetings is to identify strategies to improve learning, where there are deficiencies, and determine the curricular, fiscal, or administrative impact of these strategies.

Phase IV. Evaluation of the improvement and closing of the appraisal cycle. The recommendations of phase III are implemented, and it is evaluated if they bring the desired result. This information is used to determine if necessary to adopt new strategies or if the strategies implemented have yielded results (evidence of improvement). If the recommendations were successful, the assessment cycle is closed; if not, we continue looking for alternatives and implementing them until the desired result is achieved (completing the cycle).

Learning occurs in different settings of the educational institution. For example, the classroom, academic programs, or student support programs, such as tutoring and counseling, exist to support student's academic success. The application of the assessment cycle involves the articulation of efforts to collect data on learning in these various settings. The dominant practice is to identify institutional mission learning. Data are then collected on what occurs in classrooms and academic programs to evidence achievement of the mission (Allen, 2004). The goal is that by aggregating classroom and educational program assessment data, information is generated about the institution's effectiveness in achieving its institutional mission.

Some authors consider learning assessment a humanistic data collection model because real-life or professional criteria assess the student. It is also known as authentic or real assessment (Liou and Hynes, 2012; Ponce, 2014b). The student is asked to demonstrate how much he or she knows and what he or she can do with his or her learning in the context of real-life or professional situations. For example, the student is presented with a social or professional status and is asked to write a paper presenting a solution. The faculty evaluates the paper using rubrics they designed to determine learning strengths and areas for improvement. For example, if the student demonstrated a critical analysis of the situation, if he/she assertively applied the knowledge, or if he/she wrote with propriety. It is called the humanistic model because its nature allows the student to personally demonstrate what he or she knows and what he or she can do with the learning, rather than responding to exams and standardized tests. The dominant instruments in this model are rubrics and portfolios. Humanistic models dominate the learning assessment scenario in North American universities. This model demands a leading role as a specialist in the subject to be assessed. The logic of the model is as follows.

Phase I. Identification of learning. The faculty and administration of the educational institution determine which learning from the mission and academic programs will be assessed. For example, for the institutional report, only those competencies in the academic programs connected to the institutional mission are assessed. This does not prevent the faculty from evaluating all the competencies of the program that they deem necessary. Instead, the assessment is done to respond to the needs of institutional improvement and accountability.

Phase II. Data Collection. The faculty and the institution's assessment staff develop the data collection instruments. Faculty collects student learning data.

Phase III. Data analysis and identification of actions to improve learning. Faculty evaluates students' work and determines how to improve learning. The administration considers the fiscal and administrative impact of the faculty's recommendations, and together they decide which are feasible to implement.

Phase IV. Evaluation of the improvement. The faculty implements the recommendations. The institutional assessment office supports the faculty in assessing the impact of the advice and determining their scope. This information is then used to formalize the knowledge as part of program improvement, institutional academic planning, or budget allocation.

Institutional Research Programs

There is no perfect model for measuring, assessing, and evaluating learning (Suskie, 2009). Learning is a complex and multidimensional human phenomenon. When examined as an institutional product and responsibility, it becomes much more complicated. Institutional research is the planning and orchestration of scientific studies on fundamental aspects that characterize accountability to agencies that accredit institutions or license their academic programs in aspects of institutional effectiveness or achievement of goals and objectives, institutional efficiency or management of finances, and educational quality and student success. Institutional research programs use quantitative, qualitative, or mixed research designs to study the many complex questions and phenomena that arise in educational institutions with learning. Determining the effectiveness of an educational institution in promoting student learning can be done with some degree of confidence if data collection on outcomes is conducted within a framework or model that allows it (Judd and Keith, 2012). In other words, the effectiveness of learning data collection is enhanced if institutional studies and learning assessment are conceptualized as part of the same integrated model. In this way, the two strategies complement each other and increase the validity of their data because they support each other. Alexander Astin presents a model of institutional learning research that provides the possibility of integrating learning assessment. His model focuses on three aspects when investigating learning in an educational institution: input, environment, and output (Astin and Linsing, 2010):

Input. The input refers to the student profile of the institution. This model begins with creating a profile of the student admitted to the institution: Who is the student? And how does he or she arrive academically at the institution? Variables that help to understand learning at the beginning of the educational experience are as follows: high-school GPA, college admission test scores, vocational interest tests, placement tests, and first-year interest and needs surveys. Astin's recommended research designs are as follows:

(a) correlation studies between data on input and output variables; (b) experimental, controlled, or natural studies to understand the process of input, environment, and output.

Environment. This refers to the academic experience or treatment received by the student. The variables that help understand the scholarly treatment are the following: the curriculum, teaching techniques, evaluation strategies, personal relationships with fellow students or their professors, and the student's level of satisfaction. For Judd and Keith (2012), this is one of the strengths of Astin's model because it makes researchers aware of the differences between institutions: What intervention does the student receive? What does that intervention look like? According to Astin and Linsing (2010), the recommended research designs for understanding the environment are qualitative studies.

Product. This refers to student learning. Astin proposes investigating learning in terms of value-added or the student's academic gain to explain how and how much the institution contributes to this development. For these purposes, the following studies are recommended: experimental studies, alumni studies, student satisfaction studies, and longitudinal cohort studies. Cohort studies are investigations of changes in groups of students. For example, let us say you want to study students admitted to a college in 2010. You identify the students and track them as a group (cohort) to identify what changes occurred in them due to their studies. Cohort studies are used for many purposes: to measure retention and graduation rates, to determine learning, to understand academic experiences, or whether they are working in professional fields related to what they studied.

Institutional research is an accountability strategy that complements learning assessment and evaluation in determining the institution's effectiveness, efficiency, and academic quality. It complements learning assessment efforts as these two strategies are intentionally linked to generating data to understand learning in an educational institution. The following studies are common in institutional research and typically support learning assessment and evaluation efforts. These are studies that seek to understand the learner as the *raison d'être* of all educational management. Common studies on students are the following:

1. **Studies on student profile and characteristics.** Studies that seek to understand demographic, psychosocial, economic, or personal aspects of students. These studies are used to create profiles of the student

population as a group or by years of research. They are also used to get to know the students attending the institution, identify their educational needs, align educational services to their educational needs, guide faculty and staff in creating sensitivity, or evaluate marketing and recruitment strategies in terms of the recruited student profile.

2. **Students' needs and interests**. Studies that seek to identify the students' motivations or life and professional aspirations. They are used to know students' values, attitudes, and beliefs as population segments of the institution. These studies serve many purposes: to understand students' values, align or develop programs and services that respond to or facilitate their aspirations and motivations, and revise institutional policies, curricula, or teaching and evaluation techniques to respond to the realities of their students.

3. **Courses with the highest failure rates.** Studies that seek to identify which are the courses where students show higher failure rates and why. There are courses in which, due to their nature and complexity, incidences of failure are common. The study of the courses with the highest failure rates is used to understand the students' academic deficiencies and the teaching strategies used by the professors who teach them and to identify alternatives to improve performance in these courses. There are occasions where failure exceeds the nature of the course due to the professors' teaching styles or the way they evaluate them.

4. **Professional revalidation studies.** Professional revalidations have become an indicator of accreditation in many disciplines of study. Professional remarks are understood as an indicator of the effectiveness of the academic program. Professional revalidation studies seek to understand all the variables that explain student behavior on these tests: content measured by the test, curricular alignment to elicit learning for the demands of the test, the teaching and assessment strategies used by the program, and the institution's historical performance on the test. These studies are used to improve student performance on these tests because the findings are used to revise curricula, identify more aligned teaching/assessment strategies, or develop programmatic assessment systems to monitor student development.

5. **Success and professional recognition of students**. Studies that seek to document the professional success stories of the educational institution's graduates. These success stories serve to understand the effect of

the institution on students. They are also used to develop marketing campaigns.

6. **Retention and dropout studies.** Studies that seek to understand the reasons, motives, or factors that influence students to persist or drop out of their education. These studies serve many purposes: to identify and explain student dropout or persistence in the context of the institution, to develop intervention programs, to evaluate the relevance of student support services in light of the findings, or to evaluate the curriculum and teaching and assessment strategies to understand their influence on learning.

7. **Studies of graduates**. Studies that seek to identify and describe the professional development of students after they graduate. These studies serve several purposes: to identify the strengths and deficiencies of academic preparation with the job market, identify impacts of the academic degree acquired on their professional mobility, and identify personal changes resulting from the academic degree obtained.

8. **Employer studies.** Studies that seek to understand the experience of employers with the students they hire from universities. These studies are developed and used for several purposes: identifying strengths and deficiencies in students in light of labor demands and improving curricula to make them more responsive to the job market, and identifying labor demands.

9. **Student satisfaction surveys.** Studies that seek to understand student satisfaction with the educational institution, study program, courses, or various aspects of the academic experience.

10. **Organizational climate studies.** Studies that seek to understand the level of satisfaction of the faculty and other employees with the institution.

11. **Labor market studies.** Studies that seek to understand the demands and needs of the labor market. The objective is to design new research programs or keep the current ones responsive to the labor market.

12. **Fiscal studies**. Studies that seek to understand the fiscal status of the institution. The most common analysis is that of revenues (income) vs. expenditures (expenses). These studies are essential to make decisions that allow and guarantee the institution's fiscal health and the fiscal impact of the findings.

13. **Market positioning studies.** Studies that seek to understand how the educational institution compares with other similar educational institutions in the market it serves. These studies indicate how the

institution compares enrollment, costs, and other criteria with other educational institutions in the region, country, or those with which it is compared. These studies are used to determine the development or growth of the institution.

14. **Institutional image studies.** Studies on how the institutional image is perceived in the eyes of the public. There are image studies based on perceptions of random samples of population groups and others that classify institutions by quality criteria. These studies are used to understand the strengths and areas for improvement that others perceive about the institution. These studies are used for marketing the institution to achieve the desired student recruitment.

15. **Employability rates.** Studies that seek to identify whether students are hired after graduation. These studies are used to align academic programs to the needs and interests of the employment market. The objective is to have educational programs that provide students with the competencies that employers are looking for and hire them.

16. **Learning in guidance, counseling, and tutoring programs**. Studies that seek to understand the effect of these programs and services on the development and academic success of the students who use them. These studies are used to understand students' academic deficiencies and develop strategies to support educational programs to achieve student success.

Strategies for Explaining Learning as an Institutional Product

A recurrent position in the literature when researching and evaluating learning as an institutional product, especially in universities, is to conceptualize it as a series of changes in students. This position explains learning in terms of changes in knowledge, values, and skills or abilities that occur in students as a consequence of their education (e.g., Allen, 2004; Austin and Linsing Antonio, 2012; Middaugh, 2010; Suskie, 2009).

Knowledge. It means handling information, facts, and definitions on various topics that constitute the professional or personal training that students pursue with their studies (Allen, 2004; Suskie, 2009). Some authors indicate that measuring cognitive changes in learning constitutes the main focus of learning assessment (Middaugh, 2010 p. 93).

Values. It means the evaluative positions of liking, acceptance, or rejection about their professional or personal training (Allen, 2004; Suskie, 2009); for example, whether you understand that statistics is an essential topic in your professional training and why.

Aptitudes (skills and abilities). It means the execution skills to exercise your profession; for example, if you can write accurately or analyze situations to propose solutions in light of the information represented. The information and professional values you acquire form the basis of your skills (Allen, 2004; Suskie, 2009).

This positioning translates into three strategies to collect data and explain learning through measurement, assessment, and evaluation. These strategies are explained below.

Value-Added Theory

It means explaining learning in terms of the changes that the educational institution "adds" to students. Astin and Linsing (2012) call this positioning the "development of talent." According to these authors, talent development in students is aligned to the educational goals pursued by educational institutions. The following quote describes this vision for purposes of learning as an institutional product:

"Under the talent development view, excellence is determined by our ability to develop the talents of our students and faculty to the fullest extent possible. The fundamental premise underlying the talent development concept is that true excellence lies in the institution's ability to affect its students and faculty favorably, to enhance their intellectual and scholarly development, and to make a positive difference in their lives. As far as educational excellence is concerned, the most excellent institutions are, in this view, those that have the greatest impact- "add the most value," as an economist would say-to the student's knowledge and personal development" (Asting and Linsing, 2012, p. 7).

The value-added theory is based on the principle of individual learner differences. Studies on human growth and development show that biological and genetic aspects largely determine to learn. Although all human beings go through similar stages of human growth and development, and these stages can be influenced by environmental factors such as education, some students will mature faster than others and reach different levels of development and learning. In other words, not all students will be outstanding, regardless of the effort made by the institution and faculty to develop them. From a value-added perspective, learning assessment means collecting data on student learning from a base or starting point (base-line) to determine the changes or value added by the institution (Allen, 2004). The value-added theory is widely accepted among faculty and staff of student support services programs for its humanistic side. It tends to be criticized and sometimes rejected among administrators, whose interest is the institution's image, because of the performance of students in standardized tests or professional revalidations. This model is implemented as follows.

Diagnosis. It means to evaluate or determine the academic and social levels with which students begin their educational process. From this data, the assessment is developed to establish the value, talent, or changes that the institution adds to the students. In many universities, this assessment is known as diagnostic and placement testing.

Development. This means collecting data on learning at strategic points in student development. At some colleges, this means completing specific courses/experiences, such as general education courses. At other colleges, it may be completing a certain number of academic program credits to appreciate the transition from "freshmen" to "sophomore" to "senior."

Exit. It means to collect data at some core point of the academic program when the student is close to graduating and completing his or her studies. This information is compared with the student's diagnostic data, and changes or value added are determined. Some universities use the same diagnostic tests to determine with certainty the changes. In other universities, portfolios are used to collect students' work throughout their studies to account for differences in maturation and learning, as evidenced by the quality and depth of academic projects.

Theory of Absolute Enforceability

It means measuring or determining whether or not the student has or does not have the knowledge and skills expected of them to graduate or be admitted to the profession they wish to practice. This positioning explains learning in terms of categories such as "masters or does not master," "does or does not satisfy," and "does or does not meet" the expected or established. Explaining learning in absolute terms entails determining the accepted level of mastery that the learner must achieve. The most common strategy for assessing the level of mastery begins by identifying the competencies demanded by the labor market and, from there, the knowledge that needs to be taught to achieve them. The professions that use this approach to learning have translated it into professional or quality standards and standardized tests. From the absolute performance model, assessment means collecting data to determine if the student meets the requirements for entry, development, and admission to the profession. Some universities use this model with internal academic success criteria established by the faculty. The goal is to use the data to guide students to careers whose requirements are consistent with their knowledge and skills. In this way, they increase the likelihood of students' academic success and the institution's retention and graduation rates. The learning model centered on the demonstration of absolute performance has gained much popularity among proponents of knowledge standardization and educational quality improvement. Some policy makers favored this approach in the U.S. Federal Department of Education and administrators because assessment is reduced to measures or test scores, which are then easily compared across groups of students, academic departments, or educational institutions. Some teachers reject this approach because it does not take into account individual student and institutional differences. Some politicians and administrators have used this model to hold faculty accountable for student learning. Expressions such as "there are not bad students but bad teachers" are common among proponents of this view. This stance has generated controversy because it does not contemplate the complex dynamics of learning and teaching processes that transcend the simple outlook that "if the student does not learn, it is because the teacher does not know how to teach." From the perspective of research and evaluation of institutional learning, it is implemented as follows.

Entrance to the study program. The student must obtain the minimum score in some established tests required by the universities as a requirement for admission. Typically, these tests emphasize language skills, mathematics, general knowledge, and critical thinking. These skills are considered the basis for specialized knowledge of the profession.

Academic development. Tests are administered to determine the student's academic development. The student must pass the test and achieve the established minimum score to continue his or her studies. The strategy and the minimum expected score are considered indicators of success when taking the admission or revalidation test to the profession.

Professional revalidation. A standardized test measures the student's mastery of knowledge and skills to determine if he/she reaches the required score to be admitted to the profession.

Theory of Cognitive Development

It means to evidence through measurements of students' cognitive development or gain (Middaugh, 2010). This type of assessment focuses on students' knowledge to determine at the end of their studies how much they learned or developed. This type of assessment focuses on standardized input, process, and output test measures to determine pre- and post-measurement comparisons of the student. These standardized measures also serve to make comparisons between institutions using the same tests. This model has been criticized in the literature because its use relies on commercial instruments and not because it responds to the objectives of the academic programs or the cultural reality of the institution (Forsythe and Keith, 2004).

The Challenges of Accountability for Learning as an Institutional Product

Accountability for learning has provoked many controversies in the United States. Six debates highlight the complexity and challenges facing educational researcher-evaluators in measuring, assessing, evaluating, evaluating, and researching learning. These challenges are discussed as they are unveiled in the research process.

Institutional Relevance of Learning Assessment Models

Assessing and measuring learning has generated controversy at all levels of education, especially in universities in the United States. The following four questions regarding rubrics or standardized tests discuss debate: Which of the models is more reliable? Which is more feasible to implement in terms of cost, time, and logistics? Which has the greatest impact on learning? Where students' strengths and deficiencies are best appreciated and have the most significant potential to improve the institution? The following are the controversies surrounding these questions.

The controversy of instruments and data quality. The assessment cycle is a systematic process of collecting and analyzing data on learning. Which of these instruments is most reliable for the assessment? The humanistic assessment model dominantly uses rubrics and portfolios as instruments for data collection and analysis. They argue that learning is too complex a phenomenon to be assessed with a test. Rubrics and portfolios provide the flexibility to capture and understand the complexity of learning in diverse settings. Standardized tests lose their validity because the student does not have the vocabulary or reading skills to understand the questions to be answered. The psychometric model uses standardized tests to measure learning. They argue that tests are more accurate and reliable than rubrics for measuring and comparing the learning that occurs in an institution. Rubrics and portfolios are exposed to the subjective interpretation of the scorers. This makes them inaccurate.

The controversy of the scope of the model to understand learning. If the assessment cycle aims to generate data to improve learning, which of these models has the tremendous potential to do so? The humanistic assessment model argues that learning is too complex to be reduced to several test scores. Many teachers are not trained to understand the statistical complexity that goes into the interpretation of standardized tests. When tests are standardized, the very process of standardization makes them general and inaccurate. Understanding learning requires the quantitative and qualitative data generated from the use of rubrics and portfolios. This makes it easier to appreciate the true strengths and areas for improvement in students and the context of real-life or professional situations. The psychometric model argues that the accuracy of the data is fundamental to making decisions about learning. Rubrics and portfolios are exposed to the subjective interpretation of the scorers. This makes them inaccurate. Data quality is also questionable when faculties are not skilled in constructing rubrics and other learning measurement and assessment instruments.

The controversy of the institutional relevance of the model. Learning does not occur in a vacuum but the cultural context of the institution. Which of these models has the most significant potential to improve the institution? Proponents of the humanistic model of assessment argue that its approaches have greater sensitivity and relevance to the educational culture of the institution. Data collection instruments are developed in light of the criteria, curricula, and educational experiences to which the student is exposed at the institution. For example, data generated from rubrics and portfolios tend to make more sense to teachers when analyzing and interpreting the information because it brings the context of learning through qualitative data.

Commonly, quantitative measurement or data is not easily translated into corrective action that can be linked to the curriculum or teaching strategies. For example, saying that the arithmetic means for the criterion "handling of test information" was a 3 out of 4 does not explain why. In this sense, the qualitative data that accompanies the quantitative data allows connecting the quantitative data with the institutional context, if the corrector's comment to qualify the three he assigned was to say that the student did not present all the necessary information because he failed to specify how to solve the situation. This data allows the faculty to go into the specific aspect of the application of knowledge. The psychometric model has this shortcoming. Standardized tests respond to idealized criteria, from groups outside the institution, of what a student should or should not master. This approach can be discriminatory to the student profile of those educational institutions whose mission is to provide educational experiences to culturally disadvantaged students. This point has been a constant controversy in the history of standardized testing. Proponents of psychometric assessment argue that standardized tests allow learning in light of real quality standards congruent with the job market. The quantitative data makes it possible to appreciate where the student is about the score required by the standard. This makes it possible to develop a real education relevant to the life and professional experiences that students will experience when they enter the world of work.

The controversy of the administrative feasibility of the model in its implementation. Implementing an evaluation model of student learning, whether it is of humanistic or psychometric approach, involves an organizational structure and a logistics of implementation. Which of these models is more feasible to implement in terms of costs, faculty training, or institutional logistics for its effective implementation? The humanistic model is more complex to implement at the beginning because it must become a process of the institution and its culture. This may involve modifications or development of institutional policies and procedures and involving faculty, administrators, and staff in student support programs. This alters the institutional routine at its inception. The visible fiscal cost of the humanistic model is tied to the cost of photocopying the assessment exercises completed by students and the measurement instruments. There is also an economic investment in the computer programs used for data analysis. It may also cost the hiring of consultants to train faculty in the initial implementation phase. The complexity of the psychometric model lies more in the construction and validation of the tests than in their administration.

The complexity in the administration of the tests lies in whether the tests are administered at a time that affects the academic development of the institution. This can create discomfort and inconvenience for some teachers and students. The other challenge is whether the tests are completed on paper or electronically. Both administrations entail logistics of a test duration time, the number of students impacted, and protocols to safeguard the procedure's validity. The costs associated with standardized tests are tied to the company that designs, administers, and analyzes the tests. These costs can also be increased by the number of times per year the tests are administered. The cost of testing is always in the thousands of dollars category for all the expertise involved in design, implementation, and remediation. Its impact on the institutional budget is always visible. The psychometric model has a more significant fiscal impact on the institution than the humanistic model.

The Need for a Common Language on Learning

According to LeCompte and Aguilera-Black Bear (2012), assessing learning simply means asking whether institutional learning goals and objectives were achieved, under what criteria or standards they were performed, and how well they were achieved. Joint exercises for practical assessment are to identify and define the learning competencies to be assessed in the institution, align curricula to identify the courses where this learning is being addressed, establish clear learning objectives for accuracy in determining achievement, and establish consensus on assessing learning. This process is relatively free of problems or controversy if institutional agreements are in place. The dispute with learning and assessment and evaluation arises when data have to be compared across institutions (accountability). The absence of a common language of learning and assessment then appears. For example, although there are codes of course validations between universities, it is impossible to determine whether the educational experience in the Ethics 101 course at one private university is the same as the experience in the Ethics 101 course at another. According to LeCompte and Aguilera-Black Bear (2012), there is little scientific research to support the range of quality standards being applied to schools and universities by accrediting agencies or professional bodies. Although the assessment and institutional research models show their scientific rigor, the validity and reliability of the data change according to the political standard with which they are evaluated.

After decades of documenting how universities fulfill their institutional missions, there is still much uncertainty about educational quality

(Palmer, 2012). In the United States of America, the accountability movement has worked on standard criteria and indicators of learning and has created national data banks to facilitate comparison among institutions. However, standardizing education remains a challenge. According to Palmer (2012), the accountability movement is in its infancy.

Linking Learning Data to Accountability

Linking assessment and learning assessment data with decision-making for institutional improvement is the most frequent signal of the accreditation process in the United States of America (Suskie, 2009; Middaugh, 2010; Ponce, 2014b). For these components to be connected, the following three objectives have to be achieved: (a) analyze data on student learning (Suskie, 2009); (b) interpret the data to convert them into information about student learning—use this information to make decisions about improving education or the educational institution (Middaugh, 2010); (c) convert the information about learning into intervention strategies, if deficiencies are identified, or ratify the educational practices being used if the data indicate academic success (Middaugh, 2010). The objective of data analysis in the assessment and evaluation of learning is to convert these data into information for decision-making that facilitates institutional improvement and accountability. For Middaugh (2010), this is facilitated by understanding the difference between data and information and the need to convert data to information:

"A cardinal tenet in communicating assessment results is the understanding that data are not, in and of themselves, equivalent to information. Data has to be massaged, manipulated, and interpreted to render them into a form of information that is readily digestible and is used for planning, decision making, and the allocation of resources" (Middaugh, 2012 p. 173).

Computer programs process, summarize, and organize data but do not interpret them. Interpretation is an exercise of the assessment and learning protagonists. Interpreting data means converting the data into information about learning strengths or deficiencies that can be used to identify actions for improvement (Middaugh, 2010). A piece of data is partial information about the learning being assessed. Interpreting data means generating an explanation of the learning being studied with the data collected. It is to identify and tell the story that the data bring (Kramer, Hanson and Olsen, 2010). The first step in interpretation is to understand, build, or generate a picture of the learning with the information communicated by the data. Hence the richness and importance of collecting quantitative and qualitative data on learning to create that understanding. The construction of this understanding is facilitated when the context from which the data is generated is understood: the learning that was assessed, the questions that were used to create the data (e.g., essays, portfolios, and oral presentation), and the instruments used to measure or determine it (e.g., rubrics or standardized tests). The act of interpreting is never limited exclusively to the data collected because it always involves one's sense of students' learning (Middaugh, 2010) and their needs (Kramer, Hanson, & Olsen, 2010).

The second step in interpretation is reflection and analysis of the data to generate the grand explanation. Common strategies to confirm the explanation are identifying dominant positions, trends, or patterns in the data. The more the data aligns with the answer we generate, the more confident we feel, and the more valid the explanation becomes. Once the data is interpreted or the grand explanation is generated, we formulate recommendations for improvement that address, respond to, or are in line with the data collected. According to Kramer, Hanson, and Olsen (2010), recommendations are nothing more than establishing the priority to address the needs and deficiencies shown by students. Interpreting data and converting it to information is always an exercise that consists of comparing the learning data against some criterion that allows passing judgment. The following standards are common in the activity of interpreting data.

Criterion-focused interpretation. Data are analyzed, and student performance is judged based on the criteria and standards established by the faculty for the course or academic program. For example, how closely does the student's, or the group's, writing resemble the course objective? The evaluation of criteria is based on the faculty's expertise.

Norm-focused interpretation. Data are analyzed, and students' performances are judged in light of the group's performance. Students' scores may be ranked by ranges and percentiles, score distributions for grading, or classified in terms of above average, average (norm), and below-average scores.

Standards-focused interpretation. Data are analyzed, and student performance is judged against professional standards of quality or standardized and idealized measures of what is considered professionally accepted or unacceptable.

Decision-making relates to the analysis of data to produce actions to be taken to intervene and improve learning. Learning can be examined in three broad categories for improvement purposes: changes in students, changes in the structure of knowledge, or changes in student support services.

Eliciting change in students. Recommended intervention actions may involve bringing about changes in students. Examples of changes in students might include bringing about more favorable attitudes toward their studies, differences in study habits, or the acquisition of new skills. Changes in students, or their family environment, are always more complex and take longer to implement and observe because we are dealing with thinking human beings. In the complexity and cultural diversity in which we live, actions and interventions with students and their families must be evaluated from ethical-moral, legal-labor, and civil rights perspectives. Provoking and observing changes in students always take more time than changing the curriculum or student support programs.

Bring about changes in the structure of learning. Changes in learning structure relate to course content, teaching techniques, or academic policies. Changes to learning structures tend to be easier to implement and take less time to impact the learner. Changes to the learning structure need to be evaluated in terms of federal policies governing student grants, program licensing conditions, institutional policies, their impact on student study time, and their fiscal impact on the institution.

Bring about changes in student support services. Retention, tutoring, and counseling programs work with the student and impact student learning. Do proposed changes to improve education have to do with student support services? Recommendations to support services should be evaluated in terms of how they bring the services closer to the needs of the students, whether the proposals are feasible for the expertise and fiscal resources allocated to these programs, or whether the recommendations can be better worked from these areas, rather than from the academic area.

In the United States of America, it has become evident that the final decision with these strategies for improving education is based on the availability of funds and the administrative and political interests pursued, and not on the scientific accuracy of the data. The claim in the United States of America is the need to rescue education from administrators and politicians to let educators do their job (Alberts, 2015; March, 2015; Phillips, 2015).

Who Makes Decisions to Improve Learning?

The objective of learning assessment and evaluation is to generate data that will inform decision-making to improve the academic quality of the educational institution. The discussion in the previous section addresses the aspects to be considered to ensure valid and reliable data. However, useful and reliable data do not guarantee good decision-making because they are two different processes. Data alone do not speak for themselves; they must be interpreted.

The interpretation of data is not an act that occurs in vacuum. It is accompanied by the lens brought by the interpreter (Provezis and Jankowski, 2012). Assessment and evaluation of learning, as strategies for accountability, involve many audiences consuming the same data. These are the constituents that the accountability process brings to the discussion of assessment data (Zenisky and Languillas, 20120):

Hearings are internal to the institution. Students, faculty, institutional administrators, institutional researchers, and appraisal directors/coordinators.

Audiences external to the institution. Accrediting agency officials, parents of students, prospective students interested in the institution, and employers.

Similarly, assessment and evaluation decision-making occurs at the following levels in light of diverse interests in learning: (a) **classroom learning** involves teachers with their students; (b) **academic program** improvement involves faculty in their respective programs with the director of assessment, deans, and institutional researchers; (c) **improvement of student support structures involves** the staff of the student support programs, with the support of the institutional assessment and research staff; (d) **institutional effectiveness** involves the Chancellor with his staff from the academic, planning, finance and budget, and student affairs areas, supported by the institutional assessment and research staff.

The following recommendations emerge in the literature on how to make decision-making a more effective process (Provezis and Jankowski, 2012; Zenisky and Languilles, 2012; Palmer, 2012).

Tiered decision-making. Assessment and evaluation decision-making should not be conceptualized as one big data decision but as a series of tiered decision-making for different purposes. For example, three levels of decision-making might be: how to improve the classroom (teachers), the academic program (program faculty), and the institution (institutional management). As Palmer (2012) points out, there are data on learning that have little relevance in determining institutional effectiveness. To expect the Chancellor of an institution to tell the faculty how to teach their area of expertise in light of assessment data is not logical or reasonable.

Enrich decision-making by analyzing data from different perspectives. Learning data should be considered as an essential source of information for decision-making. This decision-making is enriched when the same data are analyzed from other optics. (a) **Learning.** Comparing the data with external criteria such as professional standards, internal comparison between groups of students, or professional revalidations. The comparison makes it possible to identify or discriminate strengths and deficiencies in the light of other students or reference groups. (b) **Institutional improvement**. Bring diverse groups and lenses to analysis and decision-making. For example, planning, finance, academic affairs, student affairs, institutional research, or assessment interpret the same data. Each of these players will bring their perspective to the data analysis, and this should produce more profitable decisions that are more responsive to the needs of the institution.

In short, decision-making is an aspect of the learning accountability process that must be planned for effectiveness (Krishman, Yin, Mahler, Lawson, Harris, & Ruedginger, 2012; Drezck McConnell & Doolittle, 2012; Banta & Pike, 2012; Palmer, 2012). As Palmer (2012) points out, educational quality remains a question mark after two decades of accountability efforts.

Drafting the Learning Assessment Report for Accountability Purposes

The report is the means of communicating results to different constituents. The following quote illustrates the role of the report in the culture of accountability:

"To demonstrate their accountability, effectiveness, and efficiency, as most higher education institutions are expected to do these days, it is not sufficient to implement good (or even very good) institutional research methods. The results must be presented in ways that are accessible, useful, and understandable [...] Ultimately, as the decision in higher education becomes increasingly more data-driven, reporting mechanisms that allow college and university administrators to draw on results efficiently will be a premium" (Zenisky and Languilles, 2012, p. 601).

There is no universal format for what the report should look like. Again, I draw on a quote from Zenisky and Languilles (2012, p. 594):

"In terms of reporting results, it seems safe to say that "one-size-fits-all" approach to assessment (and ultimately reporting assessment data) is complicated in higher education given the numerous institutional types ranging from public research universities to private, not-for-profit colleges, to community colleges (and, correspondingly, a multitude of institutional-specific missions). However, regardless of the assessment purpose, a unit of analysis under consideration, or type of evaluative instruments used, the current economic and social climate is such that all institutions share the common need for reporting their assessment results and doing so in ways that are accessible, useful, and understandable to intended audiences."

According to Provezis and Jankowski (2012), Zenisky and Languilles (2012), Palmer (2012), and Middaugh (2010), the report should consider aspects such as the following:

1. Whether the report takes the form of a written or electronic document, it must be written considering the various audiences that will receive it (Provezis and Jankowski, 2012), what use the document will have, and what information it will communicate. The audience defines the report (Zenisky and Languilles, 2012).

2. The report should be short and straightforward. It should communicate the history of the institution with the appraisal. It should use text and graphics to communicate the message (Middaugh, 2010). It can be organized by sections to respond specifically to its various audiences. It should include e-mail addresses, telephone numbers, and names of contact persons in case someone wants to learn more about the history of the institution (Provezis and Jankowski, 2012).

3. It should communicate the positive and negative aspects identified in the appraisal process. It should communicate the actions taken with the data to improve the institution (Provezis and Jankowski, 2012). The discussion should always take place in the context of the institutional mission (Judd and Keith, 2012). Not all appraisal information has to be communicated (Ponce, 2014b).

4. The report must go through a scrutiny process for validation before submitting or posting on the institution's website. The draft report must be reviewed by a committee representative of the various constituents of the institution: faculty, administrators, appraisal or institutional research staff, and students. This exercise serves to identify clarity,

inconsistencies in content, or relevance to its potential audiences (Zenisky and Lenguilles, 2012).

Evolution in the Search for Scientific Effectiveness

The research-evaluation of learning is one of the methodological challenges of educational research since its beginning at the beginning of the 20th century. At the beginning of the 21st century, the level of sophistication of educational research to determine learning is evident. Three strategies dominate in determining learning in educational institutions: standardized testing programs, learning assessment programs, and institutional research programs. Each of these programs is based on different notions of learning. At the beginning of the 21st century, it is necessary to continue studying and understanding each of these approaches to refine them. It is also imperative that the three approaches be integrated as models for collecting data on learning to improve the effectiveness of learning assessment research in educational institutions.

References

Abebe, T. & Bessell, S. (2014). Advancing ethical research with children: critical reflections on ethical guidelines. *Children's Geographies*, 12(1), 126–133.

Alberts, B. (2015). Empowering Our Best Teachers: Essential for Producing More Effective Systems of Education in the United States. In Feuer, M. J., Berman, A. I., & Atkinson, R. C. (Eds). *Past as Prologue: The National Academy of Education at 50. Members Reflect* (pp. 143–150). Washington, D.C.: National Academy of Education.

Aragón, L. (2015). *Evaluación psicológica.* 2nd Edition. Madrid: Editorial Manual Moderno.

Arias-Valencia, S & Peñaranda, F (2015). La investigación éticamente reflexionada. *Revista de la Facultad de Salud Pública de la Universidad de Antioquia*, 33(3), 444–451.

Arzola Franco, D.M. (2017). Evaluación, pruebas estandarizadas y procesos formativos, experiencias en escuelas secundarias del norte de México. *Educación*, 26(50), 28–46.

Astin, A. W. & Linsing Antonio, A. (2012). *Assessment for Excellence*, 2nd Ed. American Council on Education. Oxford: Rowman & Littlefield Publisher, Inc.

Bakiri, N. (2016). Technology and Teacher Education: A Brief Glimpse of the Research and Practice that Have Shaped the Field. *TechTrends*, 60, 21–29.

Balnaves, M. & Caputi, P. (2001). Introduction to Quantitative Research Methods: An investigative approach. London and Thousand Oaks: Sage Publications.

Barnett, R. & Parry, G. (2014). Policy Analysis Research in Higher Education: Negotiating Dilemmas. *Magis: Revista Internacional de Investigación en Educación*, 7(14), 69–84.

Barhouse Walters, P., Lareau, A. & Ranis, S. H. (2009). *Education Research on Trial: Policy Reform and the Call of Scientific Rigor.* New York and London: Routledge.

Beltrão de Lucena Córdula, E (2015). Fenomenología versus Positivismo Científico: Metodologías aplicadas a las investigaciones en las Comunidades Humanas. *Revista Intersaberes*, 10(21), 660–675.

Berliner, D. (2002). Educational Research: The Hardest Science of All. Educational *Researcher*, 31(8) 18–20

Bergman, M. (Ed).(2008). *Advances in Mixed Methods Research*. Los Angeles: Sage.

Biesta, G. J. & Burbules, N. C. (2003). *Pragmatism and Educational Research*. Oxford: Rowman & LittleField Publisher Inc. Lanham, MD.

Biesta, G. (2015). On the Two Cultures of Educational Research, and How We Might Move Ahead: Reconsidering the Ontology, Axiology and Praxeology of Education. *European Educational Research Journal*, 14 (1), 11–22.

Boddy, J. (2014). Research across cultures, within countries: Hidden ethics tensions in research with children and families? *Progress in Development Studies,* 14 (1), 91–103.

Bogdan, R. & Biklen, S. (1992). *Qualitative Research for Education: An Introduction to Theor and Methods* (2nd Ed). Boston: Allyn and Bacon.

Bokova, I (2010). Education for All: Raising to the Challenge. *UN Chronicle*, 3, 3–12.

Braster, S. (2011). The people, the poor, and the oppressed: the concept of popular education through time. *Paedagogica Historica*. 47, 1–14.

Brodisky, B. (1999). Top Ten Educational Events of the 20th Century. *The Educational Digest*. April, 4-7.

Campos, A. (2009). *Métodos mixtos de investigación: integración de la investigación cuantitativa y la investigación cualitativa*. Bogotá: Investigar el Magisterio

Campos, B. (2011). *Mejorar la práctica educative: herramienta para optimizer el rendimiento de los alumnos*. Madrid: Walters Kluwer.

Carretero, M. (2011). *Constructivismo y educación*. Buenos Aires: Paidós.

Carr, W. & Kemmis, S (1986). *Becoming Critical: Education Knowledge and Action Research*. London: Taylor & Francis Groups.

Carneiro, R. (2015). Learning: The Treasure within — Prospects for Education in the 21st Century. *European Journal of Education*, 50, 1, 101–112

Caruth, G. D. (2013). Demystifying Mixed methods research Design: A Review of the Literature. *Melvana International Journal of Education*, 3(2), 112–122.

Charles, C. M. (1988). *Introduction to Educational Research*. New York & London: Longman.

Cheung, A. C. & Slavin, R. (2016). How Methodological Features Affect Effect Sizes in Education. *Educational Researcher*, 45 (5), 283–292.

Cirino Gerena, G. (1984). *Introducción al desarrollo de pruebas escritas*. Madrid: Editorial Bohío.

Clark, C. (2011). Education(al) Research, Educational Policy-Making and Practice. *Journal of Philosophy of Education*, 45 (1), 37–57.

Coben, D. & Llorente, J. C. (2003) Conceptualising Education for All in Latin America. *Compare,* 33(1). doi: 10.1080/0305792032000035290

Coburn. C. E. & Penuel, W. R. (2016). Research–Practice Partnerships in Education: Outcomes, Dynamics, and Open Questions. *Educational Researcher*, 45(1), 48–54.

Cochran-Smith, M., Ell, F., Grudnoff, L., Ludlow, L., Haigh, M. & Hill. M. (2014). When Complexity Theory Meets Critical Realism: A Platform for Research on Initial Teacher Education. *Teacher Education Quartely*, Winter, 105–112.

Cohen, D. K. & Barnes, C.A. (1999). Research and the Purpose of Education. In E. Condliffe & L. Shulman (Eds). *Issues in Education Research*. San Francisco: Jossey-Bass Publishers.

Cohen, L. & Manion, L. (1980). *Research Methods in Education* (2nd Ed.). Sydney: Croom Helm.

Colella, F. & Díaz-Salazar, R. (2015). El Discurso de la calidad educativa: un análisis crítico. *Educación y Educadores*, 18 (2), 287–303.

Creswell, J. (2007). *Qualitative Inquiry & Research Design: Choosing Among Five Approaches* (2nd Ed). Thounsand Oaks: Sage Publications.

Creswell, J. W. & Garrett, A. L. (2008). The "Movement" of Mixed Methods Research and the Role of Educators: *South African Journal of Education*, 28, 321–333.

Creswell, J. W. (2012). *Qualitative Inquiry and Research Design: Choosing Among Five Approaches*. Thousand Oaks, CA: Sage.

Creswell, J. W. (2016). Reflections on the MMIRA. The Future of Mixed Methods Task Force Report. *Journal of Mixed Methods Research*, 10 (3), 215–219.

Cronbach, L. J. & Snow, R. E. (1977). *Aptitudes and Instructional Methods: A Handbook for Research on Interactions*. New York: Irvington Publishers.

Cooley, A. (2013). Qualitative Research in Education: The Origins, Debates, and Politics of Creating Knowledge. *Educational Studies*, 49, 247–262.

Condliffe, E. (2000). *An Elusive Science: The Troubling History of Education Research*. Chicago: The University of Chicago Press.

Condliffe, E., & Shulman, L.S. (Ed) (1999). *Issues in Education Research: Problems and Possibilities*. San Francisco: National Academy of Education and Jossey-Bass Publishers.

Cumming, B. (2012). Revisiting Philosophical and Theoretical Debates in Contemporary Educational Research and maJor Epistemological and Ontological Underpinnings. *ERIC Number*: ED537463. Online Submission.

Davis, M. & Holcombe, S. (2010). 'Whose ethics?': Codifying and enacting ethics in research settings. *Australian Aboriginal Studies*, 2, 1–10.

Denzin, N. (2009). *Qualitative Inquiry Under Fire: Toward a New Paradigm Dialogue*. Walnut Creek, CA: Left Cost Press Inc.

Denzin, N. & Lincoln, Y. (Eds.). (2008). *Collecting and Interpreting Qualitative Materials*. Thousand Oaks, CA: Sage.

Diko, N. & Bantwini, B.D. (2013). Research Politics: Some Issues in Conducting Research for the Government as a Client. *Perspective in Education*. 31(4), 15–26.

Eisenhart, M. (2015). Legislating the Value of Educational Research. In M. J. Feuer, A. I., Berman & R. C. Atkinson, (Eds). *Past as Prologue: The National Academy of Education at 50. Members Reflect* (pp. 89–94). Washington, D.C.: National Academy of Education.

Eisner, E. & Peshking., A. (Eds.) (1990). *Qualitative Inquiry in Education: The Continuing Debate*. New York: Teachers College, Columbia University.

Elliot, J. (2007). Making Evidence-Based Practice Educational. In M. Hammersley (Ed). *Educational Research and Evidence-Based Practice*. Los Angeles, London, New Delhi and Singapur: Sage Publications.

Ellis, A. (2005). What Research Is of Most Worth? In VV.AA. *Research on Educational Innovations* (4th Ed.), New York: Inc. Larchmont.

Erickson, F. (2011). A History of Qualitative Inquiry in Social and Educational Research. In N. K. Denzin & Y. S. Lincoln, (2011). *The Sage Handbook of Qualitative Research*. Sage: Los Angeles.

Feinberg, W. (2012). Critical Pragmatist and the Reconnection of Science and Values in Educational Research. *European Journal of Pragmatism and American Philosophy*, 4 (1), 222–240.

Fernández Ramírez, B. (2014). En Defensa del Relativismo: Notas Críticas desde una Posición Construccionista. *Aposta, Revista de Ciencias Sociales*, 60, 1–36.

Fernández Navas, M. Alcaraz Salarirche, N., & Sola Fernández, M. (2017). Evaluación y Pruebas Estandarizadas: Una Reflexión sobre el Sentido, Utilidad y Efectos de estas Pruebas en el Campo Educativo. *Revista Iberoamericana de Evaluación Educativa*, 10(1), 51–67.

Flick, U. (2016). Challenges for a New Critical Qualitative Inquiry: Introduction to the Special Issue. *Qualitative Inquiry*, 23(1), 1–5.

Flyvbjerg, B. (2011). Case Study. In N. K. Denzin, & Y. S. Lincoln (Ed). *The Sage Handbook of Qualitative Research*. London: Sage.

Galán, A., Ruiz-Corbella, M. & Sanchez Mellado, J. C. (2014). Repensar la Investigación Educativa: De las Relaciones Lineales al Paradigma de la Complejidad. *Revista Española de Pedagogía*, 258, 281–298.

Garcia Rupaya, C. (2012). Experiencias y repercusión de una formación en ética de investigación. *Acta Bioethica*, 18 (1), 77–81

Gee, J. P. (2015). The National Academy of Education: The Center, the Margins, and the Future. In M. J. Feuer, A. I. Berman & R. C. Atkinson, (Ed). *Past as Prologue: The National Academy of Education at 50. Members Reflect* (pp. 15–18). Washington, D.C.: National Academy of Education.

Gergen, K. J., (2013). Qualitative Inquiry and the Challenge of Scientific Status. In Giardina, M.D., & Densin, N. K., *Global Dimensions of Qualitative Research Inquiry* (pp. 29–45). Walnut Creek, C.A.: Routhledge.

Gil-Cantero, F. & Reyero, D. (2014). La Prioridad de la Filosofía de la Educación sobre las Disciplinas Empíricas de la Investigación Educativa. *Revista Española de Pedagogía*, 258, 263–280.

Glass, G.V. (2016). One Hundred Years of Research: Prudent Aspirations. *Educational Researcher*, 45(2), 69–72.

Gómez Galán, J. (Ed.). (2016). *Educational Research in Higher Education: Methods and Experiences*. Aalborg: River Publishers.

Green, B. (2010). Knowledge, the Future, and Education(al) Research: A New-Millennial Challenge. *The Australian Educational Researcher*, 37(4), 43–62.

Greene, J. (2007). *Mixed Methods in Social Inquiry*. San Francisco: Jossey-Bass & Wiley.

Guba, E. (1990). *The Paradigm Dialog*. London: Newbury Park

Guerriero ICZ, Castaño-Pineda Y (2015). La ética en las investigaciones en ciencias sociales y humanas. *Rev. Fac. Nac. Salud Pública*, 33(supl 1), S124–S127.

Gutiérrez, K. D. & Penuel, W. R. (2014). Relevance to Practice as a Criterion for Rigor. *Educational Researcher*, 43(1), 19–23.

Hammersley, M. (Ed) (2007). *Educational Research and Evidence-Based Practice*. Los Angeles, London and New Delhi: Sage Publications.

Hammersley, M. (2008). *Questioning Qualitative Inquiry: Critical Essays*. Los Angeles: Sage Publications.

Hansson, S. O. (2011). Do we Need a Special Ethics for Research? *Science Eng Ethics*, 17, 21–29.

Hargreaves, D. (2007). Teaching as research-based profession: Possibilities and Prospects. In M. Hammersley (ed). *Educational Research and Evidence-Based Practice* (pp. 3–17). Los Angeles: Sage Publications.

Hedges, L & Hanis-Martin, J. (2009). Can Non-randomized Studies Provided Evidence of Causal Effects? A Case Study Using the Regression Discontinuity Design. In P. B. Walter, A. Lareau, & S. H. Ranis (Ed.). *Education Research on Trial: Policy Reform and the Call for Scientific Rigor*. New York: Taylor & Francis Group.

Herman, J.L. & Galan, S. (1993). The Effects of Standardized Testing on Teaching and Schools. *Educational Measurement, Issues and Practice* 12, 20–25.

Hernández Madrigal, M., Ramírez Flores, E. & Gamboa Cerda, S. (2018). La Implementación de una Evaluación Estandarizada en una Institución de Educación Superior. *Innovación Educativa,* 18(76), 149–170.

Horner, J. & Minifie, F. (2011). Research Ethics I: Responsible Conduct of Research (RCR) - Historical and Contemporary Issues Pertaining to Human and Animal Experimentation. *Journal of Speech, Language, and Hearing Research*, 54, 303–329.

Johannigmeier, E. V. & Richardson, T. (2008). *Educational Research, the National Agenda, and Educational Reform: A History. A Volume in Studies in the History of Education. Information.* Charlotte, NA: Age Publishing, Inc.

Johnson, B. (2001). Toward a New Classification of Nonexperimental Quantitative Research. *Educational Researcher*, 30 (2), 3–13.

Jokonya, O. (2016). The Significance of Mixed Methods Research in Information Systems Research. *MWAIS 2016 Proceedings*. Paper 20. Retrieved from http://aisel.aisnet.org/ mwais2016/20

Kaplan, D. (2015). The Future of Quantitative Inquiry in Education: Challenges and Opportunities. In M. J. Feuer, A. I. Berman, & R. C. Atkinson (Ed.). *Past as Prologue: The National Academy of Education at*

50. Members Reflect. (pp. 109–117). Washington, D.C.: National Academy of Education.

Kilty K. M. (2015). Fifty Years Later: Access to Education as an Avenue out of Poverty. *Journal of Poverty*, 19, 324–329.

Klees, S. J (2017). Will We Achieve Education For All and Sustainable Development Goal? *Comparative Educational Review*, 61(2), 425–440.

Klemke, E.D., Hollinger, R. Rudge, D. W. & Kline, D. (ed) (1998). *Introductory Readings in the Philosophy of Science* (3rd Ed). New York: Prometheus Books.

Kincheloe, J. L. (2003) *Teachers as Researchers: Qualitative Inquiry as a Path to Empowerment*. London: Routledge.

Koishiro, M. (2013). Cultivating the Ground for the Study of Education as an Inter-disciplinary Enterprise: A Philosophical Perspective. *Educational Studies in Japan: International Yearbook*, 7 (3), 37–49.

Labaree, D (2004). *The Trouble with Ed Schools*. New Haven: Yale University Press.

Latorre, A. (2008). *La Investigación-Acción: Conocer y Cambiar la Práctica Educativa*. Barcelona: Editorial Grao.

LeCompte, M. & Aguilera-Black Bear, D. (2012). Revisiting Reliability and Validity in Higher Education Research and Program Evaluation. In C. Secolsky & B. Denison (Ed.). *Handbook on Measurement, Assessment, and Evaluation in Higher Education*. New York and London: Sage.

Lee, A. (2010). What Count as Educational Research? Spaces, Boundaries, and Alliances. *The Australian Educational Researcher*, 37(4) 63–78.

Lichtman, M. (2006). *Qualitative Research in Education: A User's Guide*. Thounsand Oaks: Sage Publications.

Lichtman, M. (Ed)(2011). *Understanding and Evaluating Qualitative Educational Research*. Thounsand Oaks: Sage Publications.

Linn, R. L. (2000). Assessment and accountability. *Educational Researcher*, 29(2), 4–16.

Lingard, B. (2015). Thinking About Theory in Educational Research: Fieldwork in Philosophy. *Educational Philosophy and Theory*, 4(2), 173–191.

Luke, A. (2011). Generalizing Across Borders: Policy and the Limits of Educational Science. *Educational Researcher*, 40(8), 367–377.

Lysenko, L., Abrami, P., Bernand, R., Degenais,C. & Janosz, M. (2014). Educational Research in Educational Practice: Predictors of Use. *Canadian Journal of Education*, 37(2), 1–26.

Manen, M. (1990). *Researching Lived Experiences: Human Science for an Action Sensitive Pedagogy*. New York: State University of New York Press.

March, J. G. (2015). Education and Don Quixote. In M. J. Feuer, A. I. Berman, & R. C. Atkinson (Ed). *Past as Prologue: The National Academy of Education at 50. Members Reflect.* (pp. 117–120). Washington, D.C.: National Academy of Education.

Martínez, L. & Prada. S. (2012). Consecuencias de Medir Progreso Educativo con Pruebas Estandarizadas: El Caso de los Estados Unidos. *Antropología y Sociología: Virajes*, 14(1), 47–63.

Mathis, W. (2003). No Child Left Behind: Costs and Benefits. *Phi Delta Kappan*, (5) 679–686.

Marks, J. (2002). A System of Comprehensive Schools or Education for All? *Education Review*, 15(1), 83–89.

Matsuura, K. (2007). Ending Poverty Through Education: The Challenge of Education for All. *UN Chronicle*, 4, 37–39.

McDonnell, L. M. (2015). Melding Political Sustainability Analysis with Education Research. In M. J. Feuer, A. I. Berman, & R. C. Atkinson (ed). *Past as Prologue: The National Academy of Education at 50. Members Reflect* (pp. 317–323). Washington, D.C.: National Academy of Education.

McDonnell, L. M. (2016). Evolving Research Perspectives on Education Politics and Policy.*Educational Researcher*, 45(2), 142–148.

McMillan, J. H. & Schumacher, S., (2005). *Investigación Educativa: Una Introducción Conceptual* (5th Ed). Madrid: Pearson and Addison

Medina, M. R. (2007). Las pruebas de aprovechamiento estandarizadas como instrumento de medición y político. *Revista Pedagogía*, 40(1), 145–187

Mejías, A. (2008). My self-as-Philosopher and My self –as– Scientist Meet to do Research in the Classroom: Some Davidsonian Notes on the Philosophy of Educational Research. *Studies in Philosophy and Education*, 27(2–3), 161–171.

Merriam, S. (2009). *Qualitative Research: A Guide to Design and Implementation*. New York: Jossey-Bass.

Mertens, D. (2005). *Research and Evaluation in Education and Psychology: Integrating Diversity with Quantitative, Qualitative, and Mixed Methods* (2nd Ed). Thousand Oaks: Sage Publications

Miles, M. & Huberman, M. (1994). *Qualitative Data Analysis: An Expanded Sourcebook* (2nd Ed). London: Sage.

Moreno, T. (2002). Cultura Profesional del Docente y Evaluación del Alumnado. *Perales Educativos*, 24 (95), 23–35.

Oakley, A. (2007). Making Evidence-Based Practice Educational. In M. Hammersley (ed). *Educational Research and Evidencie-based Practice*. Los Angeles: Sage Publications.

O'Connel, A. A. & Gray, D. L. (2011). Cause and Event: Supporting Causal Claims through Logistic Models. *Educational Psychology Review*, 23(2), 245–261

O'Leary, Z. (2004). *The Essential Guide to Doing Research*. London: Sage Publications.

Palmer, J. (2012). The Perennial Challenges of Accountability. In C. Secolsky & B. Denison (Ed.). *Handbook on Measurement, Assessment, and Evaluation in Higher Education* (pp. 57–70). New York: Routledge.

Paoletti, I. (2014). Ethics and the Social Dimension of Research Activities. *Humans Studies*, 37, 257–277.

Parsella. M., Amblerb, T. & Jacenyik-Trawogerc, C. (2014). Ethics in Higher Education Research. *Studies in Higher Education*, 39(19), 166–179.

Paul, J. (2005). *Introduction to the Philosophies of Research and Criticism in Education and the Social Sciences*. Upper Sanddle River: Pearson.

Pendlebury, S. & Eslin, P. (2001). Representation, Identification and trust: Towards and Ethics of Educational Research. *Journal of Philosophy of Education*, 35(3), 387–406

Peñalva, J. (2014). The Non-theoretical View of Educational Theory: Scientific, Epistemological, and Methodological Assumptions. *Journal of Philosophy of Education*, 48 (3), 400–415.

Peters, M. (2012). Editorial. Educational Research and the Philosophy of Context. Educational *Philosophy and Theory*, 44(8), 793–800.

Peters, S (2007). Education for All? A Historical Analysis of International Inclusive Education Policy and Individual with Disabilities. *Journal of Disability Policy Studies*, 18(2) 98–108.

Pless, M (2014) Stories from the Margins of the Educational system. *Journal of Youth Studies*, 17(2), 236–251. doi: 10.1080/13676261.2013. 815700.

Phillips, D.C. & Burbules, N. (2000). *Postpositivism and Educational Research*. New York: Rowman & Littlefield Publishers, Inc.

Phillips, D.C. (2005). The Contested Nature of Empirical Educational Research (and Why Philosophy of Education Offers Little Help). *Journal of Philosophy of Education*, 39 (4), 577–597.

Phillips, D.C. (2009). A Quixotic Quest? Philosophical Issues in Assessing the Quality of Educational Research. In P. Walters, A. Lareau & S. Ranis (Ed.). *Education Research on Trial: Policy Reform and the Call for Scientific Rigor*. New York: Routledge.

Phillips, D. C. (2014). Research in the Hard Sciences, and in Very Hard "Softer" Domains. *Educational Researcher*, 43(1), 9–11.

Phillips, D. C. (2015). Philosophy and the National Academy of Education. In M. J. Feuer, A. I., Berman & R. C. Atkinson (Ed). *Past as Prologue: The National Academy of Education at 50. Members Reflect* (pp. 121–126). Washington, D.C.: National Academy of Education.

Ponce, O. A. (2011). *Investigación de Métodos Mixtos en Educación: Filosofía y Metodología*. Hato Rey: Publicaciones Puertorriqueñas Inc.

Ponce, O.A. (2014a) *Investigación Cualitativa en Educación: Teoría, Prácticas y Debates*. San Juan: Publicaciones Puertorriqueñas Inc.

Ponce, O. A. (2014b). *Avalúo del Aprendizaje y Calidad Educativa*: Teorías, Prácticas. San Juan: Debates. Publicaciones Puertorriqueñas Inc.

Ponce, O. A. (2014c). *Investigación de Métodos Mixtos en Educación*. Hato Rey: Publicaciones Puertorriqueñas Inc.

Ponce, O. A. & Pagán-Maldonado, N. (2015). Mixed Methods Research for Education: Capturing the Complexity of the Profession. *International Journal of Educational Excellence, 1(1), 111–135.

Ponce, O. A. & Pagán-Maldonado, N. (2016). Investigación Educativa: Retos y Oportunidades. In J. Gómez Galán, E. López Meneses & L. Molina (Eds.). *Research Foundations of the Social Sciences* (pp. 110–121). Cupey: UMET Press.

Ponce, O. A. (2016). *Investigación Educativa*. San Juan: Publicaciones Puertorriqueñas Inc.

Ponce, O. A. & Pagán-Maldonado, N. (2016). Investigación Educativa: Retos y Oportunidades. In J. Gómez Galán, E. López Meneses & L. Molina (Eds.). *Research Foundations of the Social Sciences* (pp. 110–121). Cupey: UMET Press.

Ponce, O. A. (2016). *Investigación Educativa*. San Juan: Publicaciones Puertorriqueñas Inc.

Ponce, O. A., Pagán-Maldonado, N. & Claudio-Campos, L. (2017). *Redacción de Informes de Investigación*. San Juan: Publicaciones Puertorriqueñas.

Ponce, O. A., Pagán-Maldonado, N. & Gómez Galán, J. (2017). *Filosofía de la Investigación Educativa en una Era Global: Retos y Oportunidades de Efectividad Científica*. San Juan: Publicaciones Puertorriqueñas, Inc.

Ponce, O. A., & Pagán-Maldonado, N. (2017). Educational Research in the 21st century: Challenges and Opportunities for Scientific Effectiveness. *International Journal of Educational Research and Innovation*, 8, 24–37.

Ponce, O. A., Gómez Galán, J., & Pagán-Maldonado, N. (2017). Philosophy of Science and Educational Research. Strategies for Scientific Effectiveness and Improvement of the Education. *European Journal of Science and Theology*, 13(4), 35–46

Ponce, O. (2017). Investigación Educativa como un Movimiento Internacional: Nuevas Fronteras. *International Journal of Educational Research and Innovation*, 8, I–IV.

Ponce, O. A., Pagán-Maldonado, N. & Gómez Galán, J. (2018). *Philosophy of Educational Research in a Global Era: Challenges and Oportunities for Scientific Effectiveness*. San Juan: Publicaciones Puertorriqueñas, Inc.

Ponce, O. A., Gómez Galán, J. & Pagán-Maldonado, N. (2018). Investigación-Evaluación en una Era de Rendición de Cuentas: Perspectiva Internacional. In VV.AA. *Experiencias Pedagógicas e Innovación Educativa: Aportaciones desde la Praxis Docente e Investigadora* (pp. 2628–2642). Barcelona: Octaedro.

Popham, J.W. (1996). Why Standardized Tests don't Measure Educational Quality? *Educational Leadership*, 56(6), 8–15.

Pring, R. (2000). *Philosophy of Educational Research* (2nd Ed). London: Continuum.

Pring, R. (2007). Reclaiming Philosophy for Educational Research. *Educational Review*, 59(3) 315–330.

Radford. M (2006). Researching Classrooms: Complexity and Chaos. *British Educational Research Journal*, 32 (2), 177–190

Ranis, S.H. (2009). Blending Quality and Utility: Lessons Learned from the Educational Research Debates. In P. B. Walter, A. Lareau, & S. H. Ranis (Ed.). *Education Research on Trial: Policy Reform and the Call for Scientific Rigor* (pp. 125–141). New York: Taylor & Francis Group.

Ravitch, S. M. (2014). The Transformative Power of Taking an Inquiry Stance on Practice: Practitioner Research as Narrative and Counter-Narrative. *Perspective of Urban Education* 11(1), 1–10.

Rice, J. A. & Vastola, M (2009). Who Needs Critical Agency?: Educational Research and the Rhetorical Economy of Globalization. *Educational Philosophy and Theory*. 43(2), 148–161.

Rogers, A. (2012). The Case for Educational Pluralism. First Things. *Journal of Religion & Public Life*, 12, 39–43.

Rowe, M. & Oltmann, C. (2016). Randomised Controlled Trials in Educational Research: Ontological and Epistemological limitations. *African Journal of Health Professions Education*, 8 (1), 6–8.

Rudolph, J. L. (2014). Why Understanding Science Matters: The IES Research Guidelines as a Case in Point. *Educational Researcher*, 43(1), 15–18.

Scheneider, J. (2014). Closing the Gap Between the University and Schoolhouse. *Phi Delta Kappan,* 96(1), 30–35.

Shepard, L. A. (2015). If We Know So Much from Research on Learning, Why Are Educational Reforms Not Successful? In M. J. Feuer, A. I. Berman, & R. C. Atkinson (Ed). *Past as Prologue: The National Academy of Education at 50. Members Reflect.* (pp. 41–52). Washington, D.C.: National Academy of Education

Segovia, J. (1997). *Investigación Educativa y Formación del Profesorado.* Madrid: Editorial Escuela Española.

Shadish, W., Cook, T. & Campell, D. (2002). Experimental and Quasi-Experimental Designs for Generalized Causal Inferences. Boston & New York: Houghton Mifflin Company.

Shavelson, R. & Towne, L (Ed) (2002). *Scientific Research in Education.* Washington, DC.: National Research Council. National Academy Press.

Shavelson, R. J. (2015). Reflections on Scientific Research in Education. In M. J. Feuer, A. I. Berman & R. C. Atkinson (Ed). *Past as Prologue: The National Academy of Education at 50. Members Reflect* (pp. 127–134). Washington, D.C.: National Academy of Education.

Smith, M. S. (2015). Systemic Problems Require Systemic Solutions. In M. J. Feuer, A. I. Berman & R. C. Atkinson (Ed). *Past as Prologue: The National Academy of Education at 50. Members Reflect* (pp. 167–173). Washington, D.C.: National Academy of Education.

Smeyers, P. (2013). Making Sense of the Legacy of Epistemology in Education and Educational Research. *Journal of Philosophy of Education*, 47(2), 311–331.

Smeyers, P. & Depaepe, M. (2008). A Method has been Found? On Educational Research and its Methodological Preoccupations. *Paedagogica Historica*, 44(6), 625–633.

Snow, C. E. (2015). Rigor and Realism: Doing Educational Science in the Real World. *Educational Researcher*, 44(9), 460–466.

Strauss, A. & Corbin, J. (1990). *Basics of Qualitative Research: Grounded Theory Procedures and Techniques.* London: Sage Publications.

Stutchburya, K. & Foxb, A. (2009). Ethics in Educational Research: Introducing a Methodological Tool for Effective Ethical Analysis. *Cambridge Journal of Education,* 39(4), 489–504.

Tashkkori, A. & Teddlie, C. (1998). *Mixed Methodology: Combining Qualitative and Quantitative Approch*es. Thounsand Oaks: Sage Publications.

Teddlie, C. & Tashakkori, A. (2009). *Foundations of Mixed Methods Research: Integrating Quantitative and Qualitative Approaches in the Social and Behavioral Sciences.* Los Angeles: Sage.

Thompson, C. (2012). Theorizing Education and Educational Research. *Studies in Philosophy and Education,* 31 (3), 239–250.

Touriñán, J. M. (2014). Dónde Está la Educación: Definir Retos y Comprender Estrategias. A Propósito de un Libro de 2014. *Revista de Investigación en Educación,* 12(1), 6–31.

Tuck, E., Guishard, M. (2013). Uncollapsing Ethics: Racialized Sciencism, Sttler Coloniality, and Ethical Framework of Decolonial Participatory Action Research. In B. J. Porfilio, C. Malott & T. M. Kress (Ed). *Challenging Status Quo Retrenchment: New Directions in Critical Research* (pp. 3–27). London: Information Age Publishing.

Verdejo, A.L. & Medina, M. (2008). *Evaluación del Aprendizaje Estudiantil.* San Juan: Editorial Isla Negra.

Vinovkis, M. (2009). A History of Efforts to Improve the Quality of Federal Educational Research: From Gardner's Task Force to the Institute of Educational Science. In P. Walter, A. Lareau & S. H. Ranis (Ed.). *Education Research on Trial: Policy Reform and the Call for Scientific Rigor* (pp. 249–258). New York: Taylor & Francis Group.

Walters, P. B. (2009). The Politics of Knowledge. In P. Walter, A. Lareau & S. H. Ranis (Eds). *Education Research on Trial: Policy Reform and the Call for Scientific Rigor.* New York: Taylor & Francis Group.

Walter, P. B., Lareau, A., & Ranis, S. H. (2009). *Education Research on Trial: Policy Reform and the Call for Scientific Rigor.* New York: Taylor & Francis Group.

Wang, J., Lin, E., Spalding, E., Clecka, C. L. & Odell, S. J. (2011). Quality Teaching and Teacher Education: A Kaleidoscope of Notions. *Journal of Teacher Education,* 62(4), 331–338.

Woods, P. (1996). *Researching the Art of Teaching. Ethnography for Educational Use*. London: Routledge.

Wu, Ni & Wu, Jia-Li (2015). Rural Disadvantaged Children's Educational Problems and Related Policies in China. *International Journal of Intelligent Technologies and Applied Statistics*, 8(3), 225–235.

Index

About the Authors

Prof. Dr. Omar A. Ponce, Ana G. Méndez University, Cupey Campus, Puerto Rico - USA. Professor at the Division of Liberal Arts, Ana G. Méndez University, Cupey Campus in Puerto Rico. Dr Ponce does research in Educational Assessment and Higher Education. Their current project is 'Philosophy of Educational Research'. He has a PhD in Education for Leisure from New York University (United States). In addition to other important academic positions, he has been the Vice-Rector of the Universidad Metropolitana of the Sistema Universitario Ana G. Méndez (with campuses in Puerto Rico, Maryland, Florida and Texas).

Prof. Dr. José Gómez Galán, University of Extremadura, Spain, and Ana G. Méndez University, Cupey Campus, Puerto Rico - USA. Professor of Theory and History of Education, University of Extremadura, Spain, and Research Professor, Ana G. Méndez University, Cupey Campus, San Juan, Puerto Rico. Visiting Professor at several universities: Oxford (UK), Minnesota (USA), other international universities, etc. His research focuses on the scientific fields of Education, Humanities and Social Sciences in news lines of interdisciplinary dialogue. He is the director of different research groups on an international scale, both in Europe and America.

Prof. Dr. Nellie Pagán-Maldonado, Ana G. Méndez University, Cupey Campus, Puerto Rico - USA. Associate Professor at the Division of Liberal Arts, Ana G. Méndez University, Cupey Campus in Puerto Rico. Member of several professional and academic organizations: American Psychological Association, Association of Psychologists of Puerto Rico, Innovagogía, DOCERE Research Group, etc. Areas of research: human learning, cognition, evaluation & effectiveness of academic programs, research methods, research mentoring, educational research and research ethics.

Prof. Dr. Angel L. Canales Encarnación, Ana G. Méndez University, Cupey Campus, Puerto Rico - USA. Associate Professor at the Division of Liberal

Arts, Ana G. Méndez University, Cupey Campus in Puerto Rico. He is Ed.D. in Curriculum and Instruction from the Interamerican University of Puerto Rico, Metropolitan Campus. He has been Dean of the College of Education at the Universidad Metropolitana of the Sistema Universitario Ana G. Méndez (with campuses in Puerto Rico, Maryland, Florida and Texas). He is an expert in different educational specialties, and among his main professional competences is the orientation of doctoral students in Education. He was Director of the Office of Assessment, Institutional Research and Statistics of the Puerto Rico Department of Public Education.